DRILL[illegible]

DATA VORTEX

Where the bits meet the bits

Dr Carlos Damski

Drilling Data Vortex:
Where the bits meet the bits

www.drillingdatavortex.com

ISBN-13: 978-0-9941642-0-9
First Edition: Printed November 2014
Published in Australia by Genesis Publishing and Services Pty Ltd

FOREWORD

In the introduction to this book, Dr. Carlos Damski incisively contends that there is insufficient use of systematic data analysis, benchmarking and planning in the drilling of wells. Even though there is consensus about the importance of data collection and analysis, the fact that few companies are doing it systematically may surprise not only the outsider but even some seasoned professionals in the oil industry.

We now have the capability of streaming real-time data from the rig to the office, simultaneously displaying several plots on a computer screen. This may give many the comforting impression that everything is under control. However, as Dr. Damski points out repeatedly, this is merely an expense which fails to deliver a return on investment. The oil industry is already investing a lot in producing and storing valuable information. To make sense, this investment should help us to better understand our successes and flaws, extract lessons that are actually learned and act on those to improve our technical, operational and managerial processes.

Data generated during drilling operations can certainly be effective as a tool to improve the decision-making process in any drilling operation. This is certainly one of the advantages of collecting real time information. Additionally, the data collected is very useful in post-drilling analyses. However, a meaningful study of past operations is one of the most problematic tasks faced by any operations department. With hectic schedules and the tendency to always look ahead and focus on the challenges of the next project, it is very hard to commit the time and expertise required to examine the huge amount of information generated by a previous well. Herein may lie the answer to the question: "Why do we repeatedly make the same mistake?"

The process of analyzing information with the goal of improving future operations must become systematic and less time consuming - independent of the size, cost or geographic location of the operation. Nowadays, the industry has tools specifically designed to help operators face this challenge. Any company can easily access help in determining the best drilling analysis methodology and the most suitable drilling analysis software package for its particular operations.

The term "drilling analysis" has been in use for some time and several articles have been written on the subject. In my view, what makes this book unique is its very didactic explanation of the steps involved in data collection, storage, quality control and analysis. Additionally, several business cases are presented showing actual industry situations where different problems were solved using the techniques described in the book.

In the recent past, real time data generated during drilling operations was used as forensic evidence to determine why major incidents occurred. In the future, let us find a better way to use the data generated in our operations. Correct implementation and use of data analysis can not only avoid repetition of those incidents but also improve operational results. As an industry we can always do better. The tool is in our hands.

JC Cunha, PhD., P.Eng.
October 2014

"Not everything that can be counted counts, and not everything that counts can be counted"
— **Albert Einstein**

PREFACE

Since I read the excellent book "Being Digital' by Nicholas Negroponte (Negroponte 1996) almost 20 years ago, I've been fascinated by the dual world interpretation of atoms and bits, or the real world and the digital world. Those are both sides of the same coin, but as a computerized version of "the map is not the territory" (after Alford Korzybski) there are gaps in the representation and in the amount of data to represent the reality. This gap is narrowing as we gain more computing power to run more complete models and with more sensors to collect more data.

This is happening across the board in all industries, including oil & gas. It is nice to see it occurring in seismic work, reservoir simulation and visualization, real-time drilling, etc. All of this provides an excellent view of what used to be a "guesstimation".

In drilling alone, there are many facets to be interpreted in the digital model. In this book I address the issues related to drilling processes undertaken with the support of data management. Moreover, my bold aim is to level the understanding between people who work in the real world and those working in the digital world, so they can easily understand each other. In some cases, the content may appear simple for one group and eye-opening to another.

My second aim is to help improve drilling processes in the real world by using different techniques calculated in the digital world. Those simulations help to devise what-if scenarios, understand risks and assess performance. All of this gives more information about what is happening (or what might happen) in the real world, but does not solve problems. Actions to fix and improve are left to technical people, as this is their core capability. The simulation just supports and enables them to ask the right questions, reducing the guess-work factor. It is important to start with the "why". I also humbly limited the scope of this book to the management level, i.e. more focused on the time and cost of drilling operations and related interventions. At the end of the day, those are the drivers of drilling campaigns. There are many other technical aspects, such as bit selection and optimization, Bottom Hole Assembly (BHA) composition, mud properties, casing design, pump pressure, bottom hole pressure, torque and drag, trajectory, etc, which need to be addressed to build and finalize the well. There is also the safety aspects to be considered along with the improvement actions. Although very important subjects, these aspects are not addressed in this book.

The correct "data management" of everything involved in drilling will help seniors to record what they learn and enable novices to learn from the past and gain insights for the future.

I tried to avoid excessive technicality and academic discussions, instead choosing to explain all concepts in a very simple way. So, hopefully, the book will be easily understood. If needed, it can always be made more complicated in a further publication. An extended discussion of the book's content and general discussion on this theme can be found at *www.drillingdatavortex.com*. This book is intended to be read by IT people, engineers, managers, business users, data analysts and everyone involved in the drilling business who are looking for ways to improve drilling processes by using data as one of their pillars.

Carlos Damski, PhD
October 2014

ACKNOWLEDGEMENTS

This book is the result of accumulated experience and innovation. As someone from an information technology (IT) background who has worked on data management in drilling for over 15 years, I have tried to summarize some personal insights gained by interacting with scientists, business people, engineers and managers on my long road to understanding drilling processes.

Along this journey I met some wonderful people and would like to thank all of them for their contribution to my experience. I would like to mention a few of them here. I might have left some out, not because I have forgotten their contribution, but for reasons of brevity.

First, my dear friend Eric Maidla, who introduced me to this field and who is a master of teaching these subjects, in addition to being a source of inspiration as a person with his strength of character and intellectual acuity. Then there is the visionary Keith Millheim who, together with Eric, initiated the innovative concept that the drilling of new wells could be made more systematic, predictable and efficient. Both of them served as my inspiration to start this journey and to carry it further.

The Commonwealth Scientific and Industrial Research Organization (CSIRO) in Australia is really an excellent place to conceive new ideas. At the early stages of these discussions I had the privilege to work with Edson Nakagawa. He is one of those rare characters who can combine sharp intellectual articulation with human compassion at the highest spiritual level. His ideas, behavior and friendship are very dear to me. I also have had countless discussions and 'trial-and-errors' with Rosemary Irrgang and Simon Kravis, who truly give the word "scientist" its full value as the driver of further advances in the technology.

Another great source of ideas is Gerhard Thonhauser. It has been a pleasure to work with him as a fountain of endless ideas, inspirational concepts and lots of fun!

Clients are always an invaluable source of feedback. Petrobras has been a leading company in deep water wells for many years. Their research centre, CENPES, is second to none. I had the privilege to work with Kazuo Miura, who has boundless experience and knows what is needed in almost every single step of drilling and completions. He has been both a source of inspirational and practical ideas and a nightmare when it comes to making them happen in the real world. He and Francisco Torres have been able to untangle those completions sequences in a neat way. Another creative brain is Augusto Borella, who is pushing the limits further with innovative ideas. Renato Pinheiro is a grand contributor with his guts in implementing the technology and demanding new requirements. In Anadarko, Willie Iyoho has managed to implement some of those concepts and formalize them further. His publications are proof of this diligent work.

All the people at Genesis do Brasil have contributed their share. In particular, Luiz Felipe Martins deserves credit. His tireless commitment to these ideas has enabled the delivery of outstanding real results. Our countless Skype™ conferences have stretched different aspects of the technology presented here.

Thanks are also due to the staff at Genesis Petroleum Technologies who actually implemented intricate software defying statistical concepts, user interfaces and the IT corporate conundrum. A special thank you to John Anderson who put up with some weird requests and made the whole thing hang together, somehow.

Most of all, I thank Anna Damski for her contribution, not only to this book, but also to my life as a whole. She is an excellent technical person, contributing to the software development, product analysis, liaison with clients, and supporting me in my quest to promote my beliefs. She is also a source of love in my life; none of this technical knowledge would make much sense without it. To her, I dedicate this book, with love.

Table of content

Disclaimer

Throughout this book, the author presented several "expert opinions" which are based on his experience in this area. The material contained in this publication is made available on the understanding that the author is not providing professional advice, and that readers exercise their own skill and care with respect to its use, and seek independent advice if necessary for their specific case. The content does not necessarily reflect the opinion of Genesis Petroleum Technologies. The author also mentions the names of several products to illustrate the text. However, this book does not intend to be a complete catalogue of products. The author did not receive any request or benefit to include or exclude product names from this text. The product names are the copyright of each vendor and it is their responsibility to fulfill the features described in this book. The author's sole purpose is to convey possible features resulting from data management and employ examples to partially illustrate those features.

The author makes no representations or warranties as to the contents or accuracy of the information contained in this publication. To the extent permitted by law, the author disclaims liability to any person or organization in respect of anything done, or not done, as a result of reliance on information contained in this publication.

Table of figures

"In God we trust, all others bring data"
W.E. Deming

1.INTRODUCTION

Welcome to the 21st century! It is nothing more than common sense that more and more companies today rely on data to run their businesses faster, more efficiently and more profitably. Oil & gas companies are no different. They are enormous organizations, spread all over the world and sub-contracting many parts of their work. If you compare oil & gas companies with their peers in the top 20 largest organizations in the world (Figure 1), it might look "easier" to run non-oil & gas companies.

Oil & gas companies make extensive use of data management "in" the business, in a similar way to other large organizations. However, compared to other industry sectors, they are still behind when it comes to using it "on" the business. How would sectors such as retail, finance, car manufacturing, and others stay competitive in their business niches if they did not place heavy reliance on their data to help improve business performance?

Rank	Company Name	Revenues (US$mm)	Profits (US$mm)
1	Wal-Mart Stores	476,294	16,022
2	Royal Dutch Shell	459,599	16,371
3	Sinopec Group	457,201	8,932
4	China National Petroleum	432,007	18,504
5	Exxon Mobil	407,666	32,580
6	BP	396,217	23,451
7	State Grid	333,386	7,982
8	Volkswagen	261,539	12,071
9	Toyota Motor	256,454	18,198
10	Glencore	232,694	-7,402
11	TOTAL	227,882	11,204
12	Chevron	220,356	21,423
13	Samsung Electronics	208,938	27,245
14	Berkshire Hathaway	182,150	19,476
15	Apple	170,910	37,037
16	AXA	165,893	5,950
17	GAZPROM	165,016	35,769
18	E.ON	162,560	2,843
19	Phillips 66	161,175	3,726
20	DAIMLER	156,628	9,083

Figure 1: Top 20 largest companies in the world by revenue from Fortune 500 in 2013

Conversely, there is no doubt about the amount and quality of high-tech equipment available for use by oil & gas companies today. Drilling a well from above 2,000 meters (6561 ft) of water and reaching a reservoir through a complex geology at 5,000 meters (16400 ft) below the seabed is a marvelous engineering achievement. No-one can underestimate the complexity and risks associated with these operations. If you were to compare companies with equivalent equipment and conditions, you would find that the main differentiator in performance would be in the process of how efficiently and effectively they use their tools.

The days of "easy wells", good reservoirs, close to production facilities, etc, are gone! The days of drilling a well based on "my 30 years of experience" does not "cut the mustard" anymore. As with any other sector in the economy, oil & gas companies must look harder into their data, construct proper data management systems, ascertain correct key performance indicators (KPIs), and use it all to improve their business outcomes. There is no other way around it.

There are many aspects to discuss about the use of data in oil & gas companies - from reservoir estimation, logging and seismic interpretation to drilling, production, refining, etc. In this book I will concentrate mainly on the drilling aspects of exploration & production (E&P) activities. Obviously, other areas are also important, but the increasing costs and complexity of drilling wells has become a major financial obstacle to exploiting oil and gas reserves.

Drilling operations are of special interest for these reasons:

- **Lack of knowledge retention** - there is a widely-known generation gap when it comes to drilling engineers. To bridge this gap, there is an urgent need to record lessons learned, describe processes, maintain field knowledge, etc;

- **Lack of systematic data analysis, benchmarking and planning** - this is a relatively new and very elusive area in drilling. Few companies are doing this systematically and not many vendors can provide a truly unbiased service for the entire well life cycle.

In this book I will try to reconcile the two intertwined, but often disconnected, areas of drilling engineering and IT. There is a popular belief that the IT department is trying to run the business. However, engineers often see the IT department as a hindrance to moving the business forward. This is similar to a blind person trying to describe an elephant, touching only some parts of it. Obviously, IT activities are ever-increasing in complexity in all levels and directions. They have their own problems while engineers have to surmount the complexity of today's drilling environment.

This epicenter of the clash of gigantic worlds led to the creation of the title of this book, which attempts to break through the "glass ceiling" and achieve understanding among IT professionals, engineers, business analysts and executives involved in this area. With this in mind, the book describes all aspects at a basic level, without being academic or verbose. The intention is to establish common ground so everyone involved can understand each other at some essential level.

Chapter 2 explains the reason for this book. Chapter 3 lays the foundation of data management. It can be simple for IT people but revealing for engineers. Chapters 4 and 5 bring it all together with practical examples in drilling. Chapter 6 gives some business cases and Chapter 7 points to not-so-future trends and cutting edge technologies.

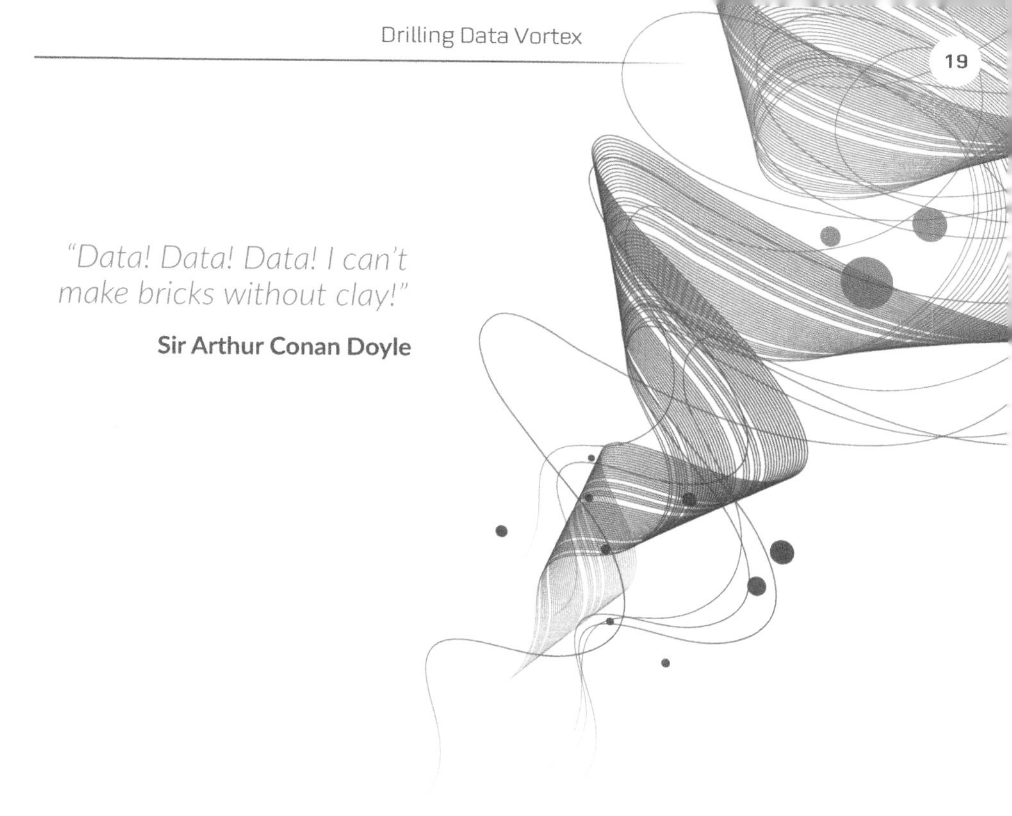

"Data! Data! Data! I can't make bricks without clay!"

Sir Arthur Conan Doyle

2. BITS ARE BITS

2.1. DATA, DATA, DATA

The term "data" is used hundreds of time throughout this book. Without being academic or formal, the term is used simply as a raw value that comes from sensors or observations. There are many definitions and structures, which can be built on top of "data", that are not very relevant to this book. In simple terms:

- **Data** are facts, numbers or individual entities without context or purpose;
- **Information** is data that has been organized into a meaningful context (to aid decision-making) readily available in documents and databases;
- **Knowledge** is the human capacity (potential and actual ability) to interpret and take effective action in varied situations.

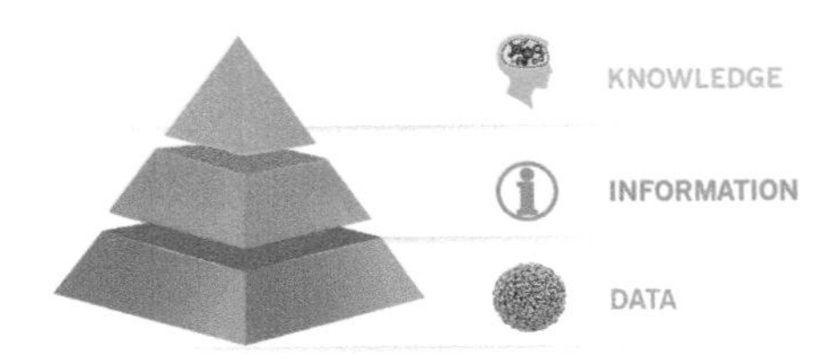

Figure 2 shows an example in drilling terms where dates and numbers are meaningless without the context of start/end time and depths. The most abstract term "knowledge" would originate from the interpretation of the information. Some of this interpretation can be automated in a computer program and some is subject to the person looking into it. Most of the time we will use "data" in a loose context that can mean any of those levels of interpretations, for simplicity and ease of reading. The purpose here is to convey the idea, without being academic.

OPERATION TIME		DEPTH	
Start Date/Time	End Date/Time	Start Meter	End Meter
10/02/2006 22.00	10/02/2006 23.00	920	920
10/02/2006 23.00	11/02/2006 4.00	920	920
11/02/2006 10.00	11/02/2006 10.00	920	920
11/02/2006 10.00	11/02/2006 20.00	920	1010
11/02/2006 20.00	22/02/2006 11.25	1010	1010
22/02/2006 11.25	22/02/2006 15.55	1010	1010

Drilling Operation

INFORMATION

DATA

KNOWLEDGE

Figure 2: Data, information and knowledge in drilling terms

The "data management" term has been in vogue for a while. More recently, as the volume of raw data increases, the term "big data" (BD) has emerged. All of this creates some confusion for the uninitiated in IT lingo and in "supernatural powers" to see the non-physical world.

There are endless publications and conferences to address "data management" issues. They are usually self-centric and look into this area as an end in itself. For any business perspective, we should look into "data management" as a means to fulfill business needs. That's the main problem with "data management projects" and "data evangelists". Most of them forget the ultimate end result, which is to deliver business intelligence and, somehow, improve the bottom line (productivity, safety, streamlining, etc) of organizations.

> *Thou shall begin with the end in mind.*

Many believe that streaming real-time mud log data from rig site to the office, and displaying it as a wiggle plot, is "data management". In an IT sense it is, as they are collecting, storing and producing some plots with it. In a business sense it is just an expense without a return on investment (ROI). It will only make sense when some decision to improve business processes can be made. In my opinion, that should be the main purpose of any "data management project". I believe that everyone involved needs to understand the "cost X value" of those projects as shown in Figure 3.

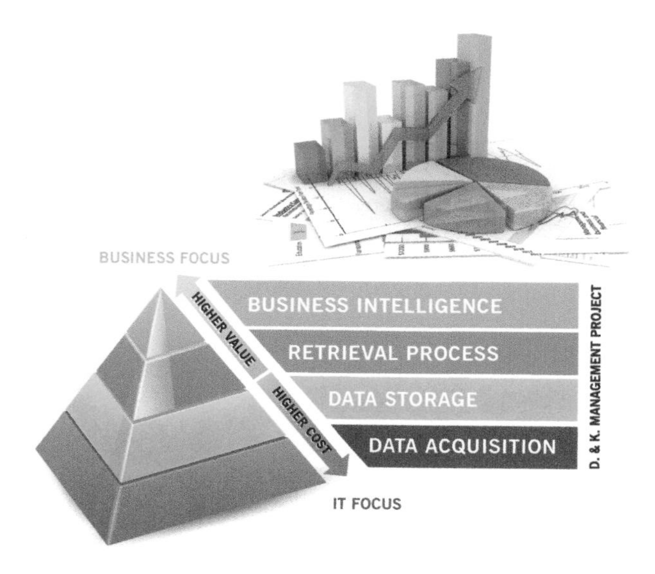

Figure 3: Data management – cost vs value in data management

Today, there is an increase in costs to collect, store and retrieve all data generated. Vendors are very creative in developing new sensors and new methods to acquire data from any possible aspect of a drilling operation. The result is a bigger IT infrastructure to support this overload. The most recent fashionable term is "big data" (BD) which basically means vendors are providing solutions designed to acquire an even larger amount of data. This is one of those weird situations where the supply exceeds the demand. Moreover, it does create a "need" to come up with ways to interpret this large amount of data, which was unknown in the first place. It goes without saying that sometimes it does indeed create some business benefits as an end result. The trend is similar in a way to when Apple Inc. invented the iPad without a specific demand being in place, and it became a huge success after being equipped with the right technologies. Such situations are, however, more the exception than the norm.

A drilling bit is a marvelous piece of steel, created after extensive experience, models and field tests, and crafted for some very specific circumstances. Vendors publish expected bit performance in a similar way to the performance information provided by car tire manufacturers. The tricky part is to actually measure its performance in the real world and statistically benchmark the same or similar bits in different wells and compare bit suppliers. The bit performance is one considerable component of rate of penetration (ROP), as this directly affects the total drilling time. This is a simple case where we collect information (bit trips, formation, wear codes, etc) about many bits and use this for business improvement. Note for drillers: have you seen the bit ROP performance statistics for 8 ½ inch bits used in the last 10 wells in the field you are working on?

It is not only the bits, but geology, casing, operations, etc, that exist in the physical and digital world. We are increasingly using digital versions more than the physical ones to understand the past and forecast future performance. As an industry deeply rooted in pieces of steel and decades of hands-on experience, it is a challenge to embrace and value the digital versions. Certainly, one does not exclude the other, and this book shows the value of the digital version.

2.2 THE ROLE OF DATA MANAGEMENT IN THE E&P BUSINESS

E&P has a high financial impact in any oil & gas company. It underpins all asset valuation and exploration, the main source of income for the company. The reservoir, rigs, facilities and pipelines are all tangibles assets that are valued and registered in the balance sheet. What becomes interesting is that no "data" has a tangible value and therefore does not appear as an asset. It would be interesting to see "120 electronic logs recorded as LAS file for well XYZ" with a value of $123,456 in the accountant's books! It does not appear as an asset because it is not tangible, but also because its interpretation is what creates the tangible asset, which has value. Data or "data management" does not therefore have value per se, but only when it is used to create a business value. There are, however, marginal data-selling businesses, with the value based exclusively on the quantity and quality of data available, regardless of its use. These companies are "data hubs" that collect, organize and sell reports of access to raw data worldwide. The cost and value of the data, among many other topics in the E&P business, are discussed by Steve Hawtin (2013) in his interesting book The Management of Oil Industry Exploration & Production Data.

If data is available then the difference between two decisions is only a matter of "data interpretation" or, in more fancy terms "business intelligence" (BI), which can generate better reservoir estimations, drilling programs, completions configuration, etc. It is this BI that can set apart the performance between two companies operating in the same field and with similar equipment. More companies are already doing "internal benchmarking" to compare how efficiently they are operating in different regions. Even better is to have "external benchmarking" where they compare their performance to that of their peers. The comparison usually leads to four quartiles of results, and everyone claims they operate in the first quartile! This is a misleading comparison, where the problem is not just the quartile, but the spread between the best and worst. If the "best" is 10 times better than the "worst", then that's very different from a situation where the best is only two times "better" than the "worst". Companies need to seek uniformity in the deliverables and later on in optimizing. One way or another, we always come back to the data used and correctly interpreted in order to make those assertions. We should revisit Figure 3 where the ultimate goal for data management is to reach BI and add value to the company, as opposed to merely a cost.

2.3 BUSINESS INTELLIGENCE

In generic terms BI can be defined as:

> *"**Business intelligence** is a set of theories, methodologies, processes, architectures, and technologies that transform raw data into meaningful and useful information for business purposes. BI can handle large amounts of information to help identify and develop new opportunities. Making use of new opportunities and implementing an effective strategy can provide a competitive market advantage and long-term stability".(Adapted from Wikipedia)*

Most examples of BI are related to data plotting. This is not a matter of "seeing is believing", but one of "seeing to understanding" as visual clues are power communicators.

The simple table in Figure 4 shows bits used in similar wells and conditions.

ROP (m/h)	Well Name	Depth Out (mt)	Metrage Drilled (m)	Rotating Time (hs)	Hole Size	Serial No.	BitTy
3.2	Trainer F	5562	68	21.25	8.5	LF4919	M
3.84	Trainer D	5062	24	6.25	12.25	16591	DS83
4.205	Trainer D	4920	246	57.51	12.25	1401787	A
4.743	Trainer F	4867	198	41.76	12.25	LF3579	MS
4.903	Trainer F	5368	76	15.50	8.5	D07CX	ATMG
5.169	Trainer D	4638	199	38.50	12.25	KX8043	M02
4.848	Trainer C	3718	80	16.50	12.25	648338	SS3

Figure 4: Bits used in similar wells

The equivalent plot is shown in Figure 5, where the ROPs are sorted with the mean value as a straight line.

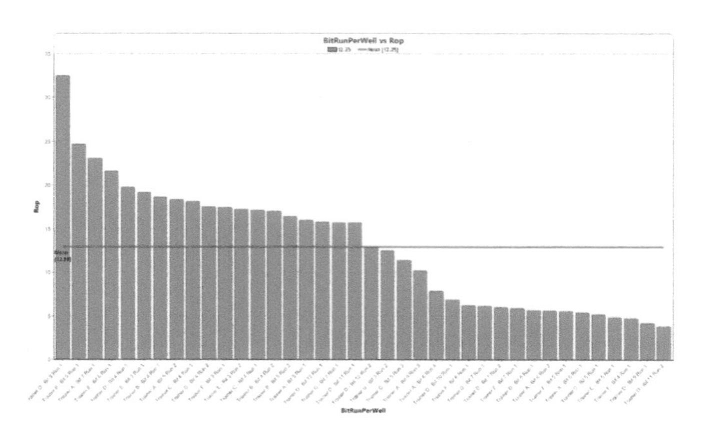

Figure 5: ROP for bit 12 ¼ used in the offset wells with the mean value

We clearly see how plots can add a new dimension to the table. That's why cross-plotting tools have become so popular among data analysts.

To arrive at those meaningful plots, there are two important steps to follow: data quality control (QC) and data aggregation. Data QC is related to:

- Correction of data outside the valid range;
- Locating and adding missing data;
- Detecting and removing (if applicable) outliers;
- Checking units and unit conversions. This is one of the most common errors, as locality is very "natural" for the user. Check meters/feet, US$/other currencies, date format, decimal point as comma, geospatial reference, etc;
- Semantic use of a coding system – needs human interpretation to decide which one is correct, which can lead to different interpretation across different persons.

The data aggregation is related to combining information in the same data source or multiple data sources. The main issues here are:

- The correct access/permission to the data – one user can have access to operational data but not to financial data;
- Cross-reference – one well has a name or identifier in one data set and is different in another data set.

All those issues are well known and will be discussed in more detail in Chapter 3. The main point here is that to reach BI no-one can underestimate those two previous steps: data QC and data aggregation. They are a MUST and need to be addressed thoroughly. They are core activities of data management and very related to IT issues.

2.4 BUSINESS BENEFITS

This book mainly discusses data management in drilling processes. However, I believe we should start with the end in mind. Although it might not be the direct responsibility of a single department to worry about the overall business, one should be aware of the main benefits and consequences of improving drilling processes in the overall business.

Chapter 4 demonstrates in more detail some ways to use drilling data. From a business perspective the most important aspects are:

- Predictability;
- Well execution follow-up;
- Benchmarking.

Predictability is related to improved ways of planning new wells. When it comes to seeking initial approval then everyone, in one way or another, plans a new well in terms of total duration and cost. The more precise and correct the sequence of operations and their duration, the better it is for the company, as it spawns decisions about investments, suppliers, resource allocation, rig management, logistics, etc. This is one of the most fundamental information areas when it comes to making financial decisions that maximize profits. Predictability is mainly achieved by correctly analyzing historical data. This is important when it comes to a single well, but it is even more so when planning an entire campaign with dozens of wells. In fact, it becomes a paramount feature for any decision-making company director. This is implemented by using correctly historical data to forecast the future operations.

Well execution follow-up is mostly related to understanding where the well is in terms of drilling process and where it is heading. Once the well is correctly predicted as above, the detailed follow-up can be used to "micro-manage" the well. A well that has completed 30% of its drilling days will already show time differences when it comes to the final duration. It will also be possible to identify performance issues related to certain operations, which can be corrected in future phases. This can be used to optimize the well in terms of reduction of non-productive time (NPT) and invisible lost time (ILT), but also to correct logistics programs with early and late deadlines for the required resources to reach the rig. This follow-up is mostly based on the timely and correct use of data acquired from the rig and matching it to the detailed plan. It is therefore basically data management that supplies all the information to optimize the well.

Benchmarking gives everyone a performance perspective. At the operational level benchmarking can compare service companies, rig crews, etc. At well level, it can compare similar wells drilled by the same rig or different rigs in the same area, as an internal benchmarking function. As external benchmarking, it can compare wells drilled by other companies using industry databases. What is most important is to "keep your eye on the ball" at your level of competency. Measuring and comparing all the time keeps you aware of where you are in relative terms. This is another fundamental business function and can be mostly implemented as a data management process.

At the end of those basic activities, among others, companies can achieve what matters most to their shareholders: maximum profitability and maximum net present value (NPV) in all projects. Data management can pave the path to success.

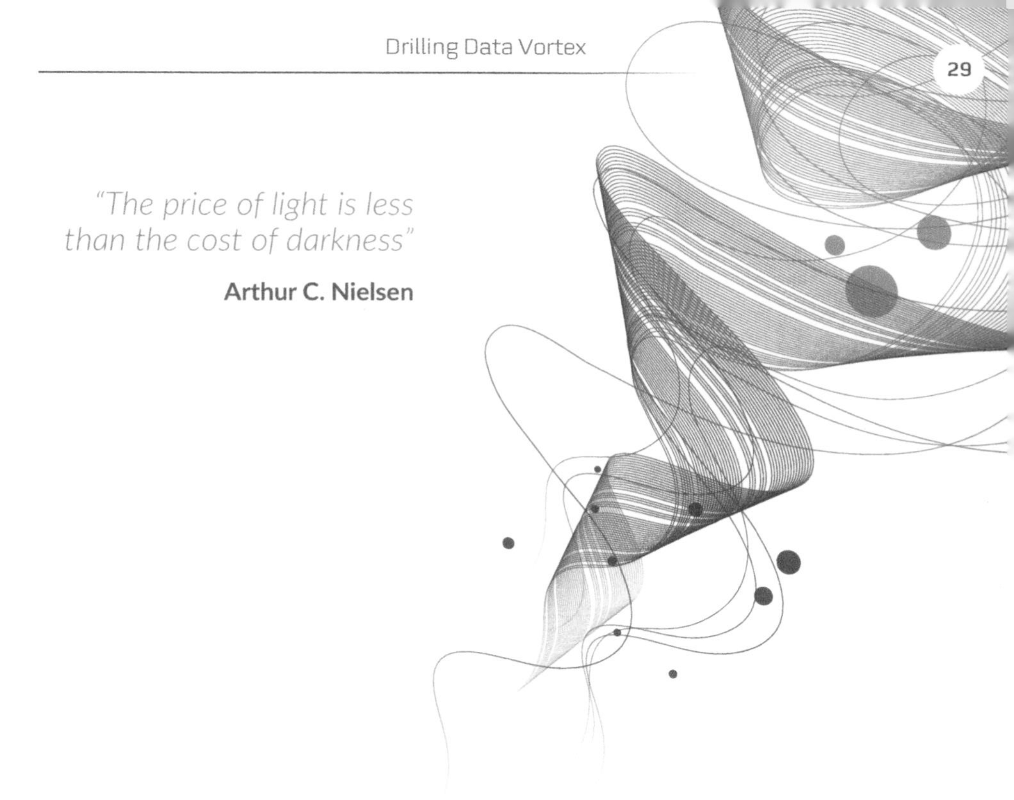

"The price of light is less than the cost of darkness"

Arthur C. Nielsen

3. DATA MANAGEMENT TECHNOLOGIES

Data management is a generic term that can cover almost anything done in IT. For a complete reference on data management applicable to all domains see DAMA International (http://www.dama.org/), which is a global association dedicated to advancing concepts and practices in this area. For the purposes of this book, data management can be summarized into five basic topics:

1. Data acquisition;
2. Data storage;
3. Data QC;
4. Data retrieval;
5. Data analysis.

Each one has its own technologies and they can be addressed and executed independently of each other.

3.1 DATA ACQUISITION

There are two forms of data sources for drilling operations: manual reporting and automatic reporting based on sensors (basically mud logs). Manual reporting, known as "daily drilling report" or "morning report system", can be as simple as a paper form faxed/emailed from the rig to the office, spreadsheet files or more sophisticated computer programs. Regardless of the method, it always involves someone in the rig manually filling a form. This is also valid for other types of data, such as geology, bit, costs, fluid used, etc, and for vendors and service providers.

Data in electronic format comes from mud loggers and sensors on the rig, which measure particular signals in different frequencies. This also includes some electronic logs performed in-between or while drilling. They are all collected in real-time and usually streamed to the office for plotting and storage. This conforms with the BD terminology, which involves the collection, storage and analyzes of a large amount of data. BD originated as a result of the increase in the volume and velocity of data collected as a consequence of reduced costs of communication and storage. Volume is due to the widespread use of sensors to measure many aspects of the drilling process and "velocity" is an increase in the speed or frequency of the data collected.

More important than the method of acquisition is the reliability of the source. Is the person filling in the form aware of the utilization of this data? Some of the people filling in those reports ask, "Why does someone need this information?" Are the sensors calibrated to provide consistent measurement? Is everyone aware of the common datum and units for all the measurements? There are many sensors collecting the same data from different service companies – which one should you rely on? Are they using the same time stamp or depth datum? Some of these issues will be discussed in the following sessions.

There are several "data brokers" who can translate and aggregate data from these data sources and move them to another format. There are also standards whereby those data brokers can import and export data such as Wellsite Information Transfer Standard Markup Language (WITSML) (http://www.energistics.org/). It is basically an XML definition to promote the exchange of well data between operators and service companies. In this way two or more companies can send and receive data from each other, even if their databases have different structures. See "Appendix B - WITSML - An Energistics Standard" for more information about WITSML.

This is just the beginning of the long journey the data has to travel until it becomes useful for decision-making in the company, as will be illustrated in the next sections. Section 4.2 addresses the reporting system in more detail in the context of the drilling process.

3.2 DATA STORAGE

Data storage is a source of many misconceptions. It is quite a clear concept if you are coming from an IT perspective, but it is not natural for engineers. All engineers are familiar with MS Excel™, which combines data with formulas in the same file, and all of this runs from the program itself. That is a nice feature but does not scale up for corporate applications. A corporate application follows the basic model shown in Figure 6.

Figure 6: A simple model of a single corporate application

In this model a "user" interacts with an "application", which is a computer program. The "application" then stores and retrieves data on a "database", which is a file residing somewhere (can be in the same computer or elsewhere). I would like to make this model very clear as I met many engineers who call, for example, a "morning report system" a "database system", which is an incorrect concept by any interpretation. The "database" is just a repository of raw data, usually organized in a set of "tables" which are similar to a spreadsheet. It is important to have such organized and separated storage so they are optimized for space allocation, speed of retrieval, data protection, back-up, etc. Normally, those databases are managed by products called database management systems (DBMS) such as Oracle, MSSQL Server, MS Access, etc. IT people are concerned with security (who has access to the application and/or database), environment version control (operational system, DBMS, etc) and the general response time of the entire system. End users are usually not concerned with those problems and sometimes underestimate the "IT issues". Please, do not undervalue the complexity of today's Windows/Unix servers/Citrix, networks, multiple DBMS and widespread use of techniques to securely access those applications.

Databases are considered "structured" data repositories as they have a formal method to organize and retrieve the data. Opposite to this, there is "unstructured" data, which is basically text data such as reports, presentations, images, etc. The unstructured data is much harder to store and retrieve as it has many different formats. Computers have a hard time to understand written texts and moreover to extract specific data within a context.

In some special cases, there is a need for special database technology to store specific types of data such as electronic logs, seismic data, real-time data, reservoir, geospatial data, etc. These are needed due to the large amount of data and different types of internal data structures.

Finally, this distinction between database and application makes it clear that the vendor is the owner of the application but the content of the database belongs to the company. Some vendors make it very difficult for users to understand and retrieve data in the database, so they can continue to license the application as the only means to access the data. It is a right for most users to challenge the vendor and ask for clear access and understanding of the content in the database, as this is a company asset. Today, there are many software tools to retrieve and use data from other applications' databases.

Another characteristic of this model is that, by separating the data from the application, it is possible to enable the access of the application to many users while they can selectively have limited access to the data. For example, everyone in one department can access operational data, and only a few others in another department. In either case, only the managers can access financial data. They are using the same application and same database, but with different rights. Another characteristic is that the application can be upgraded to incorporate new features, but the database remains the same. Now it is noticeable that spreadsheets combine these two concepts; hence the difficulty of maintaining a uniform data storage, sharing, back-up, etc.

The model in Figure 6 shows a neat solution for the initial problem, but its proliferation creates other issues that most companies face today, as shown in Figure 7.

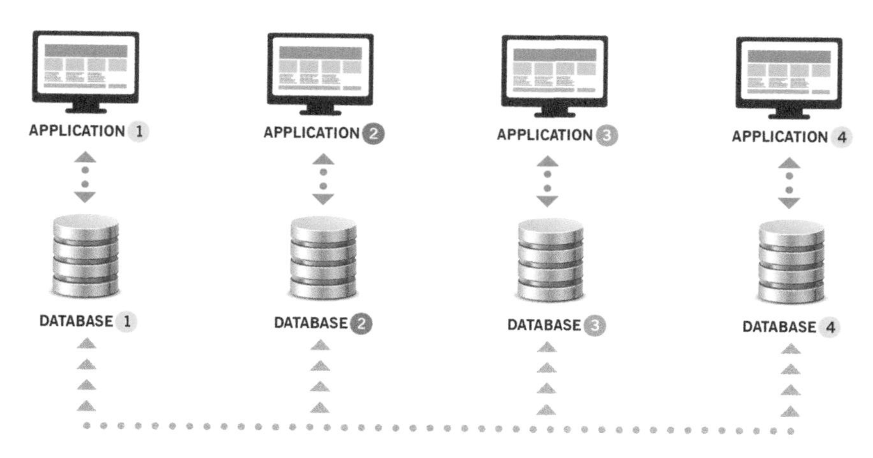

Figure 7: A model of a few corporate applications

As the number of applications grows, so does the number of databases. Each vendor wants to work with their own database. This is normally called "data silos" as they store data but do not share with other applications. As an example, just imagine the "drilling data" stored in one database, the costs related to the well in another database and the production data in a third database. One simple problem is that (usually) there is no unique common well name (or well ID) across these databases. This makes it difficult to consolidate data and create a simple report with four components: "well name", "total days drilled", "total actual cost" and "total oil produced until today". As a simple exercise, try to produce this report with the data within your company. There are some valid exceptions to this, when data needs to be processed for a specific purpose while keeping the original version elsewhere.

To address this complex problem, IT vendors came up with some solutions, which can be summarized below.

Solution 1: One database to *"rule them all"*

In this case, the proponent has a simple answer where all the information is stored in a single, corporate database, as shown in Figure 8. This would avoid the problem of "well name" and finding related information, etc. This holistic, but naive, solution falls short as vendors do not have the flexibility to adapt their application to use the corporate database. No single vendor can provide all solutions over one database, although some are trying to sell this idea.

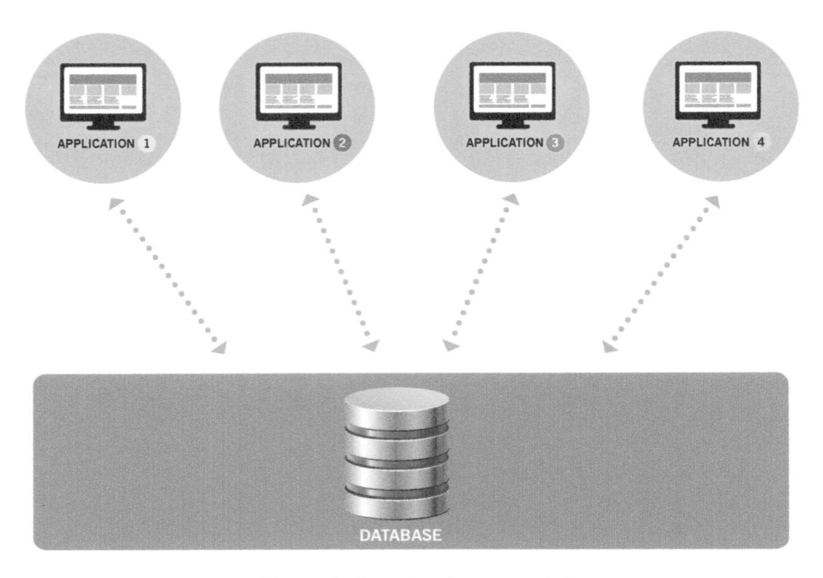

Figure 8: One database model

Solution 2: Connectors, mapping and magic wands

This is a more practical approach where everyone recognizes that the world is not perfect, but we should be able to transfer and access data using some common standards. This is more like those power point adaptors, which can be used for traveling across many countries.

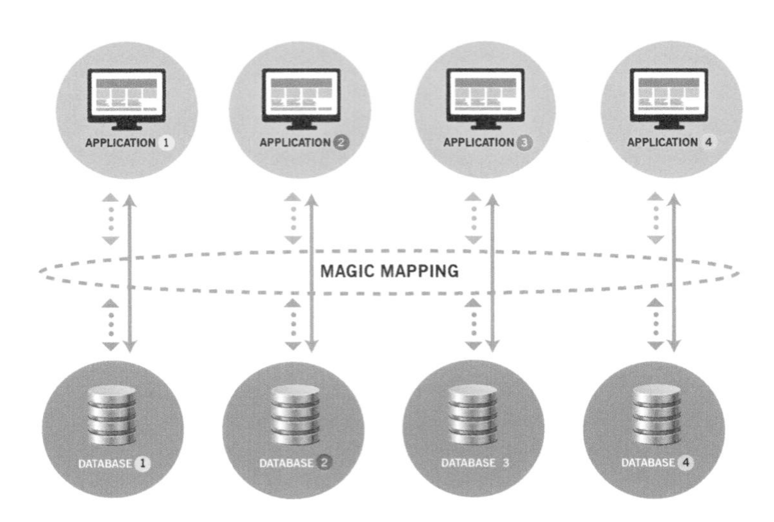

Figure 9: Database mapping

This makes more sense and it is possible to build some specific models for specific purposes. In this case, the original application continues to serve its own database but can also access or share other databases, as shown in Figure 9. More specific software components can be used, such as OpenSpirit™, and many other vendors can either develop or sell plug-ins for well-known vendors' databases. These are computer software technologies that can work in specific cases. Most of them are used to read databases and few allow writing to the database. The reason for that is the damage one can do in writing to other vendors' databases, which can affect their application's results. Although this facilitates the access and exchange of data, it does not solve the well name (well ID) issue across databases.

Solution 3: Combine data from different sources

Strictly speaking, reading information from databases for reporting purposes is now well solved with "generic" BI applications. Those are special applications that do not rely on their own database but have a number of ways to access other vendors' databases. These applications are commonplace now and usually cater for enterprise wide application, integrating many departments to provide a "single view" of the data, as shown in Figure 10.

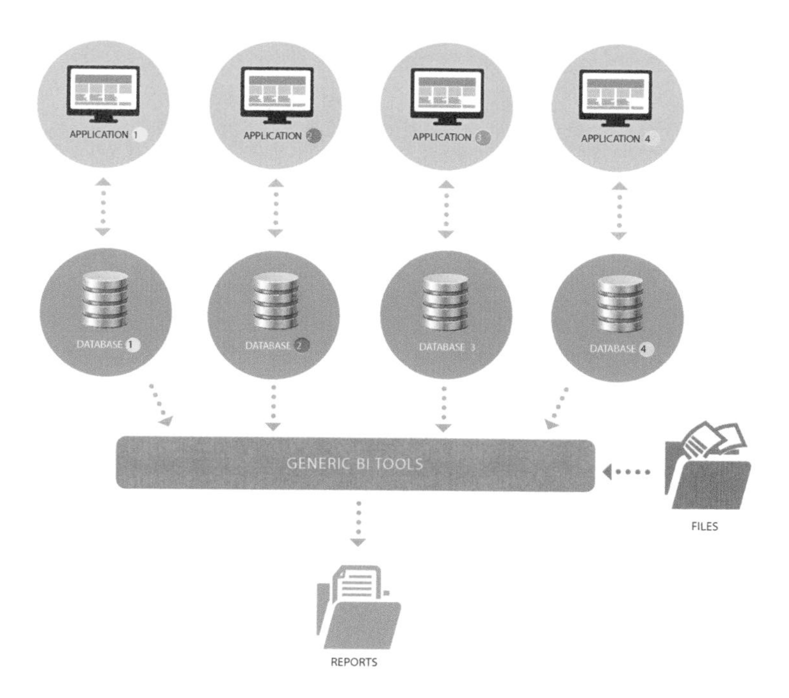

Figure 10: Generic BI tools

Examples of such BI applications are Business Objects™, Hyperion™, Cognos™, etc. This type of reporting solution still faces the problems mentioned initially as most of them have to rely on what was recorded by the applications in the databases. Another problem is the complexity of these systems. Most of them have to be used by well-trained business analysts who can understand the data structure from those databases and also the business requirements to produce the report. In this case there is a need for an ongoing dependency on the particular software platform and those specialized consultants, as there are constant changes in the business requirements and changes in the underlying databases. None of these systems are in reach of drilling engineers who need to correlate things like:

> *"The most significant trouble types occurred within a particular formation in the last five wells in this field."*

It will be good if engineers would have this information available for the next morning's meeting. However, the reality today is – "ask IT to produce the report, and receive it weeks later" – a situation that everyone has already experienced in some shape or form. Obviously this requirement is not a priority for IT. Besides having many other problems to solve, it is not IT's role to understand the nitty-gritty of the drilling business. In this case, there is a need to empower engineers to access the performance of simple queries by themselves with the use of clever software with built-in analytics specific to drilling. Most of the common engineers' requirements can be fulfilled with such a system without assistance from IT. This is the new trend in "BI application", which is much simpler than traditional applications and comes with built-in domain expertise. Examples of such systems are Spotfire™ and Genesis iVAC.

One very common mistake is to use a spreadsheet for this purpose. A spreadsheet is a powerful tool and very easy to use, and medium users can develop a quite comprehensive solution using it. There is nothing wrong with that. The problem arises with the proliferation of personal files with no standard across the company. A simple example is when the engineer asks IT (again!) for an extract from the database "to do some analysis". Once this data is received, the engineer starts "preparing the data" which means QC, fixing wrong values, changing codes, removing outliers, etc. Then the data is aggregated using their unique preference. After a substantial amount of work it is ready to be presented. While presenting the masterpiece report someone else questions the offset wells used and requires another study with a different dataset. Then, the engineer goes back to square one, asking IT for some more data, etc, etc. This is an enormous re-work and not everyone is willing to do this on a weekly basis to analyze "what-if" scenarios. In addition, another engineer has been asked to produce similar work in another field, using their own criteria as they do not know what the other has done. When the two reports arrive at the common boss's desk they are not comparable because, for example, they use a different aggregation process.

Without going any further into this data conundrum it becomes clear to everyone that using local separated files and processes, based on personal references, can lead to massive confusion and a loss of productivity. The data needs to be live from the main database and the reporting process common to all users involved.

The solution is to replace the spreadsheet in this function with a corporate application, which can read data live from databases and allow data combination/aggregation and cross-plotting facilities on this data.

3.3 DATA QUALITY CONTROL

Data QC is like cleaning the toilet. Everyone wants to use a clean toilet but no-one likes to clean it! The reason is simple. Like the sign in a shared toilet, "clean after yourself for next guest", it would involve minimal work if everyone did their share of work in the first place (including sensors). But that is not the case as not everyone does it. Imagine a rig supervisor finishing their shift and wanting to crash in bed for a deserved rest, and then finding the daily drilling report system does not allow them to save the data from the last 12 hours because the "bit wear code" is not valid!?! This infuriates the user and understandably may force them to ditch the application altogether. In cases like this most manual data entry applications relax many constraints during the data input process in the name of "user-friendliness". If it is not required, then the user may never enter it. Once the missing data is not saved to the database it goes undetected, until a few years later when someone wants to perform a study on how bits were worn out while drilling some particular formations. A quick search in the database finds nothing, even though the data entry application has a field to enter this data. After the frustration has subsided, it becomes clear that this data can never be recovered. Bummer!

That's not the end of bad news. Even if the data has been entered, this does not mean it is correct. Most common errors in drilling databases are:

- Wrong datum/unit;
- Inconsistency;
- Dispersed data source;
- Outliers/anomalies;
- Bad relationship between data entities.

Most of those cases are self-explanatory. Others are very tricky, as you can have many survey files from different runs, with some of them taken from a datum reference different to what you would have expected. Even worse, one engineer can code a particular operation in one well and another code the same operation differently in another well, making comparison very difficult.

Most people underestimate the importance of QC. Acquiring data is relatively easy and most vendors have a solution for it, but the process of quality controlling the data to make it useful for decision-making is a much longer shot. There are basically three ways people can deal with QC:

- At input process;
- While saving to the main database;
- A separated QC process.

QC done at the input process is similar to what was described earlier. The application constrains the user input data at the moment of input, for example, a value is outside the valid range, the code used is not valid, etc. As discussed, the application is usually more relaxed and allows non-QC data to be saved in the local or temporary database. Regardless of the input data, some companies adopted a more severe way to QC the data, which is when the data is saved to the main database. They use a technology called "store procedure (or database trigger)", which is basically a very small program that is executed when the data is inserted, changed or deleted from the database itself. Some advocate this solution because it is independent of the application, which is the program writing the data to the database. The QC constraint is tied to the database itself. Although it seems to be a good idea, this technology has many problems such as:

- The stored procedure itself is quite complex to maintain;
- The same code is applicable to all records in the table;
- It has technical issues with the update of new versions of the database and even more complex issues to migrate to another DBMS.

Both solutions pose some limitations from a data management perspective. QC is an evolving practice and needs to adapt to new requirements from the business. In case we need to add a new QC rule to input the "bit wear code" value we would need to change the application, which vendors can answer typically in 6-12 months for a new version. Or we could hope that the "store procedure" can handle the new rule and that someone is able to implement it, which requires back compatibility (all previously stored data needs to conform with the new rule). Both cases are quite difficult to implement and maintain.

This leads to the third option, where the QC process happens outside the database and is independent of the application. It is a process in itself. In this option, a specific application is designed to QC the data in any database. It offers a report of faulty data that can be fixed by the QC officer using the application responsible for the data and consulting other sources if necessary. This is a much more modern approach to the problem, as the QC rules can be modified by the average user, i.e. QC officer. In case we need to add the new rule for "bit wear code", the user just defines the rule in a high level language and runs the rule over partial or total records of the database to assess the situation. If later on this rule needs to be modified, the QC officer changes and runs the QC again to find other faults without the need to change the original application or the database. Moreover, the QC rules are the intellectual property of the company, not the software vendor. There are a number of these generic QC applications, such as DataVera™ and Genesis Xcheck, depending on the complexity of the environment in which they will be used. For example, with Genesis Xcheck, the user can define QC rules graphically, like designing a flowchart, making it very easy to understand, as shown in Figure 11.

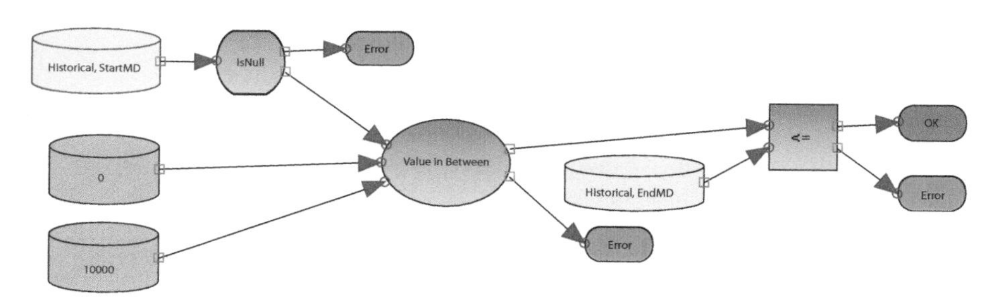

Figure 11: Defining QC rules graphically

This example shows a typical rule to compare the start and end Measured Depth (MD) for a given operation while drilling. We assume that every operation must have these two values. The "start MD" is checked to be "null", i.e. no valid data. If there is no data, then it already shows an error. If there is valid data, then this value should be between 0 and 10,000 (meters). If the value is outside the valid range, then it will show another error. If the value is within the range, the final comparison is with "end MD". The assumption is that the start MD is equal or less then end MD. If this is the case, then the record "passed" the QC, otherwise it gives an error. I can already understand the reader finding some loopholes in this rule, which can be more specific, using the same principle. It does not take long for most users to understand this process.

I advocate the last solution because it creates a process to QC, with clearly defined rules, a group responsible and the business reason for the QC, considering that the rules are tailor-made for the company's needs. It also can be used to QC near input time. In the case of a daily drilling report system, where data is entered using relaxed rules, the QC officer or an automatic process can run the report just after the data has been entered and point out the faults. In this case the user enters the data and saves it as usual. The QC rules run after the fact and the data can still be recovered if detected and found within the next day or so.

The initial problem is to find and recover the missing data months after it is entered, which is much harder, if not impossible.

Regardless of the way companies do data QC, some of them fall into a trap of saving the "good data" in another database and keeping the un-QCed data in the original database, "just in case we need the original data". This is a source of potential problems as new data continues to be fed into the un-QCed database and people will continue to use it, as it is the "most updated" in addition to using the QCed one as the "most reliable". You can already see how this potentially can create more harm than benefit for the company. The simplest way is to keep one version of the database and the QC process definitive and not to leave previous versions of data behind.

3.4 DATA RETRIEVAL

Once the data is stored in some databases, the next challenge is how to find the data you need and how to retrieve it in a useful way. To find the data some vendors use the technology called "metadata" or "data dictionary", which is basically a sophisticated index or a description of where things are and how to get them. This all depends on a wide variety of parameters and can be quite complex and also beyond the reach of an average engineer – it is really in the realm of IT people.

Another problem arises from data formatting, aggregation and combination. Most of the drilling data are organized in databases, as discussed in section 3.2. The data in the database are retrieved using a language called structured query language (SQL) which is a neat thing, but not as important for end-users as they are for IT. Without a proper construction of SQL statement syntax, it is not possible to use data efficiently from databases. It also limits what it is possible to do. The problem is not only SQL, but the underlying "data structure" used in the database. For example, suppose you want to know what the ROP is of a particular bit in a specific formation. You have a table with "bit records" where you have all the information about the bit, depths drilled, wear codes, etc, and you need to correlate it with the formation that it drilled through. As the "formation data" is stored in another table, in the same or another database, it becomes an interesting problem as a bit can drill through several formations and one formation might be drilled by more than one bit. All of this is quite "simple" if there is an application that already presents this information. The challenge is that there is always some new insight or way to look for data that the application does not already cater for. To fill this gap there are other types of applications that can retrieve data and plot in a more generic way.

It is important for users to increase usage of generic applications as this will reduce your dependence on the IT department and technology vendors, both of which can cost you time and money. These applications vary in form and complexity. They can be as simple as a graphical interface, which hides the SQL complexity, to a high-end query builder.

An example of a graphical interface is shown in Figure 12.

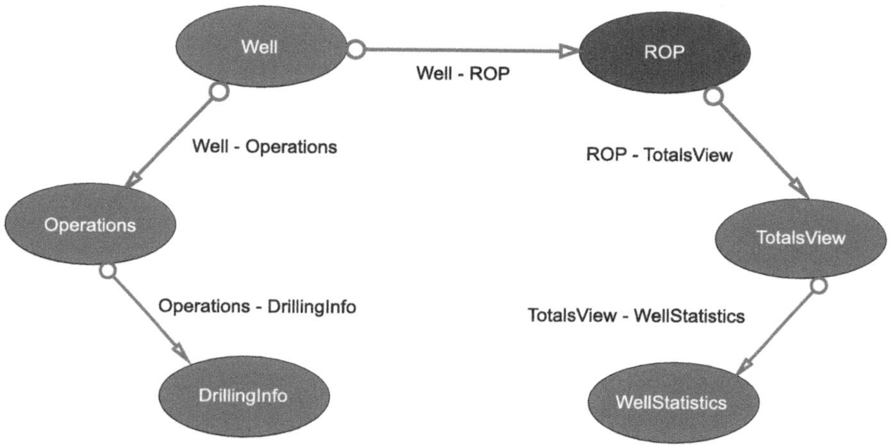

Figure 12: Graphical data preparation

In this example, a bubble represents data from one or more databases, which can be combined, filtered and manipulated before it can be used for a plot or report. This activity is similar to what users do today using Excel spreadsheets, but this is live data direct from databases, not disconnected in local machines. Once the data is prepared, it now can be used for analysis, which is discussed in the next section.

3.5 DATA ANALYSIS

Finally we reach the last phase of data management. All the data is entered, stored, QCed, and retrieved and these are the means by which we reach the data analytics stage. If the data is not analyzed with a view to helping the business to make better decisions, then it is worthless.

But what is data analysis? In a nutshell it can be described as:

> ***Data analysis*** *is a set of methods to inspect, transform and model data aiming to detect patterns, develop explanations, test hypotheses, highlighting useful information, suggest conclusions and support decision-making. (Adapted from Wikipedia)*

This is a quite a broad definition, but in reality it is a very elusive activity. The main objective must be an answer to the question: "How will this outcome help the business?" The decision-making support derived from data analytics are delivered in the form of benchmarking, KPIs comparisons: planned-vs-executed and past-vs-present, and so on. This is where the end user is empowered to ask the right questions based on solid data. It should also be possible to play what-if scenarios in order to choose the best course of action.

One important point is to keep data flowing and to add value to the next stage. In temporal terms data can reflect the past, present or future. The "present" can be this minute, this hour, this day, etc, and these terms need to be defined in relation to what is been measured and the purpose of it. This divides analytics into three basic purposes:

- **Past** – historical data are used in reporting and analysis, which basically demonstrate what happened, when it happened, why it happened and how it compares with other things that are also in the past. Most reported KPIs and benchmarking fall into this category;
- **Present** – this is related to operations and reflects what is happening now. As we receive real-time data from the rig we can analyze "instant ROP", "rig status now", etc;
- **Future** – these are predictive or pre-emptive analyzes for those events that might happen. For example, this can be related to total time to drill a new well, the potential risk of having a trouble event, etc.

For regular operations, users need all three of them. The past is used as a historical basis to understand past performance, to plan future wells and to forecast future problems. It also helps in taking decisions in the present, as problems tend to repeat themselves in similar (current) situations. There is also a need to combine historical and real-time data to provide a full picture of the events.

Some analysts might intend to use "discovery" (usually related to "data mining") which can identify trends in a larger amount of data than was initially intended. Although this is a good practice in retail, scientific experiments, etc, it has limited results in the drilling domain as most "findings" are already known beforehand. It just needs to be measured and compared.

One important aspect of data analysis is the delivery mechanism, i.e. reports, plots, etc. Plotting tools, in particular cross-plotting tools, are useful for data analysis processing and presentation. Data visualization is a very extensive field of study. For more information see the book Now You Can See It (Few, 2009). For our purposes, the important features in a good visualization tool include:

- Select any attribute to plot in X-axis and Y-axis;
- Enable multiple Y-axis;
- Enable use of bars, stackables, colour controlled, combined with trend lines, etc;
- Use data from different data sources;
- Allow trend lines (including learning curve);
- Display statistical values out of the data set;
- Allow multi-dimensions such as size, colour, shape, etc, of a data point related to attribute values;
- Enable plotting template to make uniform plots across different data sets;
- Enable web-based dashboards for end users.

Examples of such systems are Spotfire™ and Genesis Xplot. Regardless of the plotting system you are using, the most important is what to plot. Chapter 5 shows very specific examples of data analyzes in drilling.

This chapter introduced several techniques for data management. Additional information can also be found on the PPDM Association website (www.ppdm.org). PPDM is a global, not-for-profit organization within the petroleum industry designed to promote professional petroleum data management through the development and dissemination of best practice, standards and educational programs.

Data management, like any other important task, should start with the end in mind. Focusing on the final user and business needs, the data analyst should construct the requirements for data acquisition, QC, retrieval, etc, to fulfill that need. Data management is not an end in itself.

3.6 DATA SECURITY

Strictly speaking, "data security" is not in "data management" as it operates "on" data. It adds another dimension to data management and is not limited to drilling data, or E&P, but affects all data and data access in a company. It becomes a specific problem for E&P due to the increased integration of rigs and other facilities in corporate networks. Once an attack happens in the website, server, computer or personal gadget it can potentially affect production facilities, a drilling rig, etc.

Once isolated, operations facilities are now an integral part of the rest of the world with fast internet connectivity, numerous autonomous systems, and many vendors with highly sophisticated IT systems and increasingly difficult to control data access. We have seen this elsewhere – users with their smart phones and tablets are receiving real-time data anywhere and anytime. It is not far into the future that they will also be able to have some control over the facilities, and when that eventuates any breach of security on those devices can potentially jeopardize the facility as well.

The major problem IT infrastructure faces is the question of "data access" – who-can-see-what and who-can-do-what. Data storage, transmission, back-up, etc, are simpler problems to solve than authenticating a real user and avoiding access by a hacker.

The emergence of the digital oilfield has been great for the industry. There are so many benefits from bringing all IT infrastructures to the facility and the integration and collaboration of personnel involved, all helping each other. Everyone in this space has access to confidential data through their computers, most of them mobiles. All this technology is great, but it will be accompanied by problems. Today, bringing down a website by denial-of-service (DoS) has its costs and disruptions, but stopping a drilling rig or production facilities will bring enormous costs and other consequences. This is in addition to the ever-present risk of leaking confidential data.

Today, all companies are increasing policies related to data/system security. But the volume of data and different ways to distribute and access it are increasing at a faster rate, making it hard for IT infrastructure to keep pace. It is a difficult problem and there are so many aspects to it that go well beyond the scope of this book.

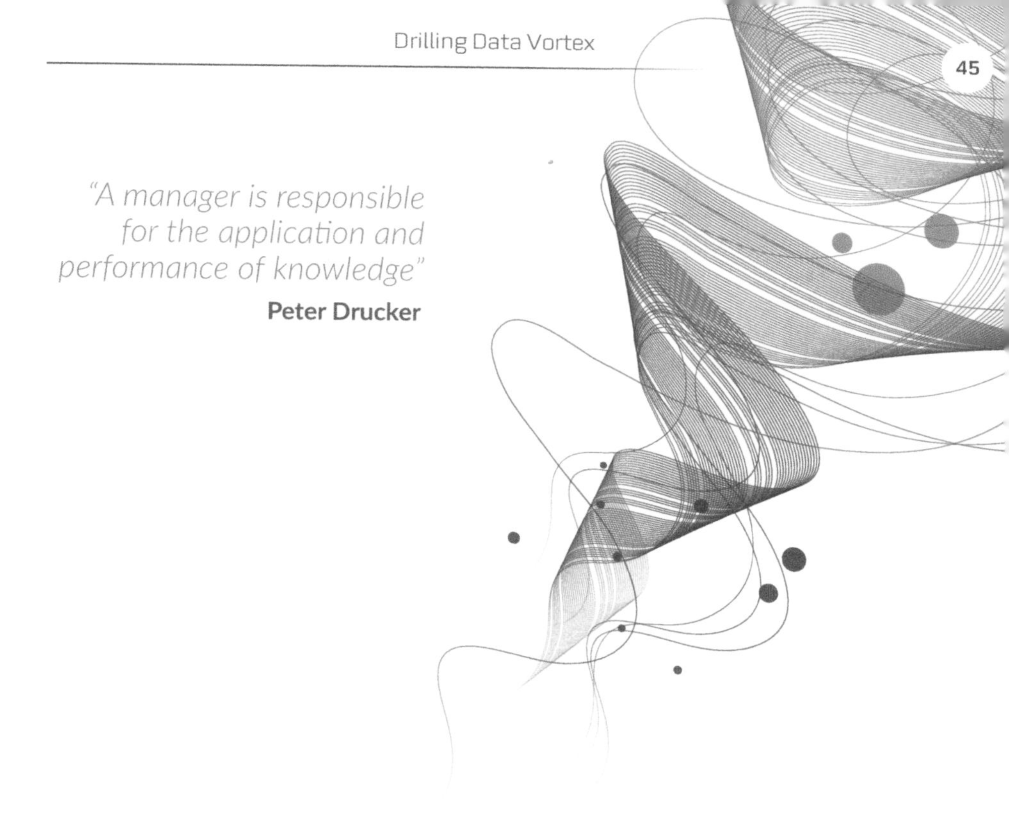

"A manager is responsible for the application and performance of knowledge"
Peter Drucker

4. USING DRILLING DATA

Chapters 2 and 3 provided the background information to finally turn drilling data into decision-making information that can help improve business processes. In a nutshell, as Peter Drucker (adapted from Lord Kelvin) would have said:

> *"You cannot improve what you do not measure."*

Not all data management is related to process improvement directly, but in this section we will place more emphasis on areas of business performance that directly affect the company's bottom line.

4.1 DRILLING PROCESS

To understand the role of drilling data in field development, it is necessary to look at a well's life cycle. One simple form is using Plan, Do, Check, Act (PDCA), which are the four basic stages of the well operations process, as shown in Figure 13. PDCA was created by W.E. Deming and is designed to lead to continuous process improvement. In this cycle companies:

- Plan the next well to drill;
- Do (execute) the well;
- Check what went right and what went wrong; and
- Act with proposed changes based on this analysis so the next well will include improvements.

Drilling data plays a fundamental role in planning and checking the process used. This chapter will dissect those two components.

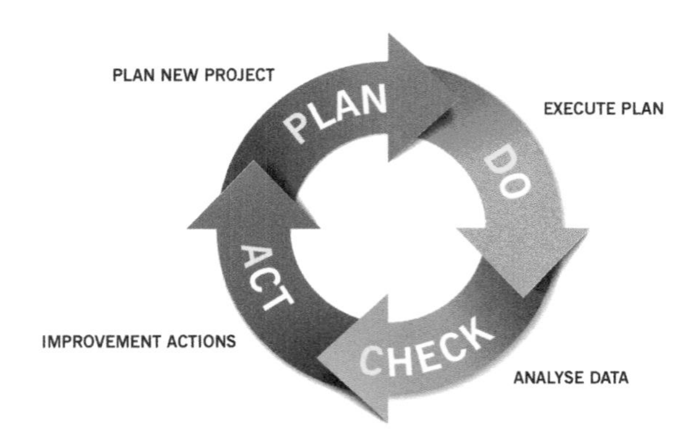

Figure 13: PDCA cycle to guarantee process improvement

Some believe PDCA stands for "Please Don't Change Anything" as they think they have already reached the best performance (or are just lazy or don't try to fix what ain't broken). This is similar to the world record in Olympics for the 100 meter sprint, where the perceived barrier of 10 seconds lasted for many years, until someone broke this limit and many others followed after that. Teams and companies behave in a similar way and the following pages will show a mechanism to break this mindset.

Companies need a better plan and a systematic way to analyze and benchmark their results. This is done by using historical data, comprehensive data preparation and statistical analysis to plan the time and cost of future wells and to properly compare different types of operations. Each intervention is different, but each one can be broken down into smaller parts with defined simpler operations that allow comparison across different types of interventions.

Once the historical data have been collected, organized and quality controlled, the engineers are ready to use it to identify where problems were experienced in the past, and where improvements may be made in planning and estimating the costs of wells. Several tools have been developed to optimize well drilling through a systematic analysis of historical data and to help engineers to investigate deeper into the data. Some of those analyzes are:

- Productive time analysis;
- Process control analysis;
- Non-productive time (NPT) analysis;
- Best composite time (BCT) - analytical technical limit;
- Bit performance analysis;
- Learning curve analysis;
- Benchmark analysis.

There are several approaches to perform this systematic analysis. One of them is the "10-step methodology" to improve overall drilling performance designed by Iyoho et al. (2004) and depicted in Figure 14. These steps are:

1. **Define project objectives and scope** in order to understand the context, identify the possible targets and design a working plan. In this section the following questions need to be answered: Where we are now? Where do we want to be? What is possible? How do we get there?
2. **Data selection and quality control/quality assurance (QC/QA).** In this step the comparable sets of data are identified, organized and compiled, and their integrity is checked in preparation for subsequent analysis;

3. **BCT analysis.** The best times for each activity from among comparable well data sets are extracted and added up to create an "ideal well" in terms of time. Then, the new planned wells will be designed in such a way that they will try to mimic the BCT. The BCT is not the technical limit in terms of what is possible, but it is in terms of what has been done;
4. **Best composite cost (BCC) analysis.** The information from the BCT is converted into costs to better analyze the financial targets - cost minimization and improved capital efficiency;
5. **Learning curve analysis.** To analyze and evaluate the performance, progress and accomplishments in general and particular areas of the drilling campaigns;
6. **Major operations analysis.** To identify NPT and problems based on a time breakdown of the major operations and drilling phases;
7. **Detailed problem analysis and opportunity identification.** To analyze and evaluate trouble-time magnitude and repetitiveness, therefore identifying improvement areas;
8. **Peer review, design changes, and technology scouting.** Study the possibility of optimizing the drilling process through changes in design, new technologies implementation, operational procedures improvements, etc;
9. **Economic benefits evaluation.** To evaluate the feasibility of the ideas proposed in the previous step;
10. **Recommendations and follow-up.** Define a series of recommendations based on the previous evaluation and continuously measure their financial impact as they are implemented, in order to continuously upgrade the BCT and BCC.

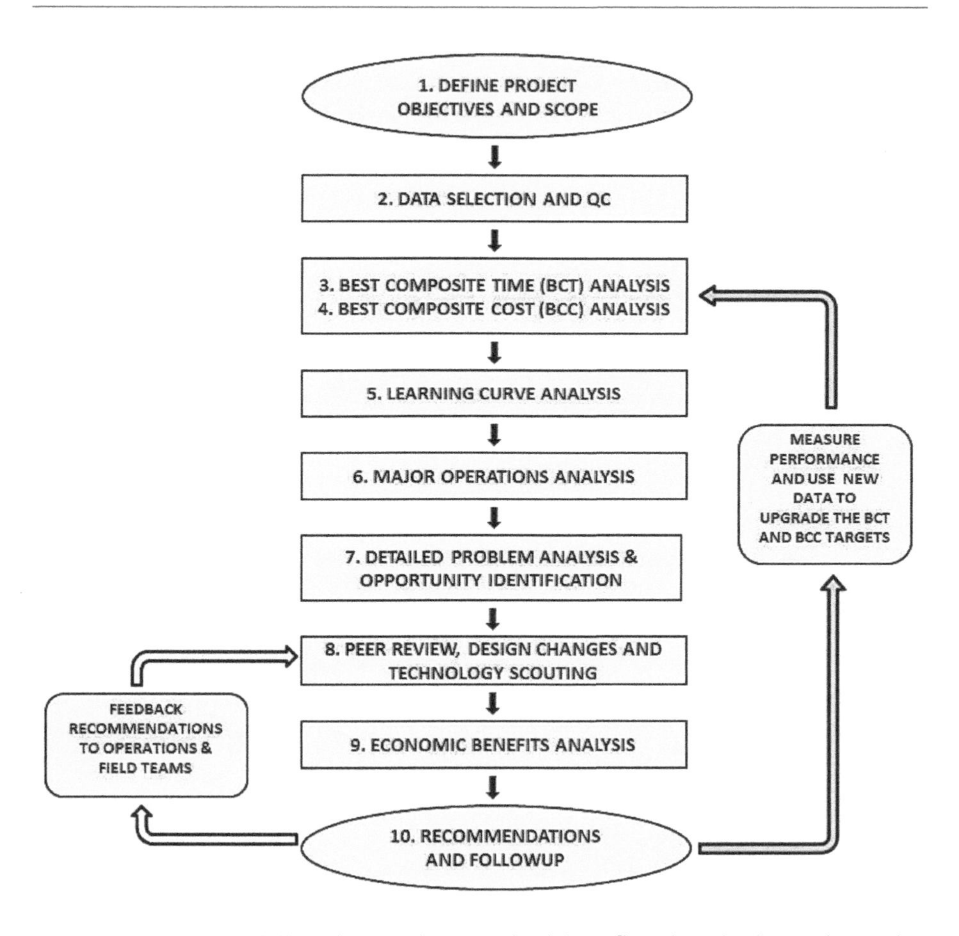

Figure 14: 10-step drilling data analysis methodology flowchart (Iyoho et al. 2004)

Without going into too many details of this process, it basically looks into improving the operational time, reducing the NPT and the "invisible lost time", as depicted in Figure 15.

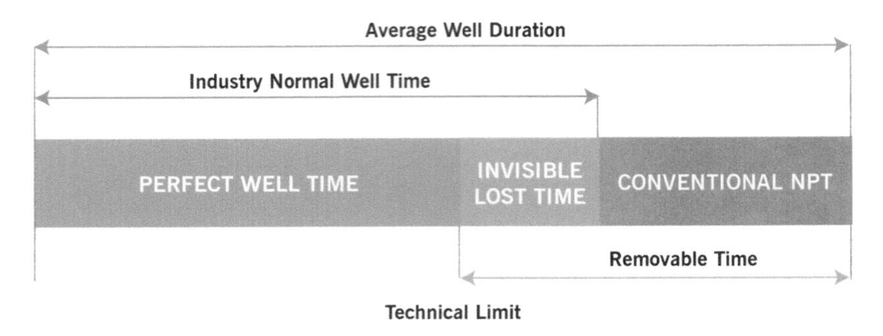

Figure 15: Relationship between the perfect well, hidden lost time, technical limit and non-productive time (Adeleye, et al. 2004)

It also assumes the continuous measurement of BCT to reach further and further optimization until it achieves the "technical limit", as shown in Figure 16.

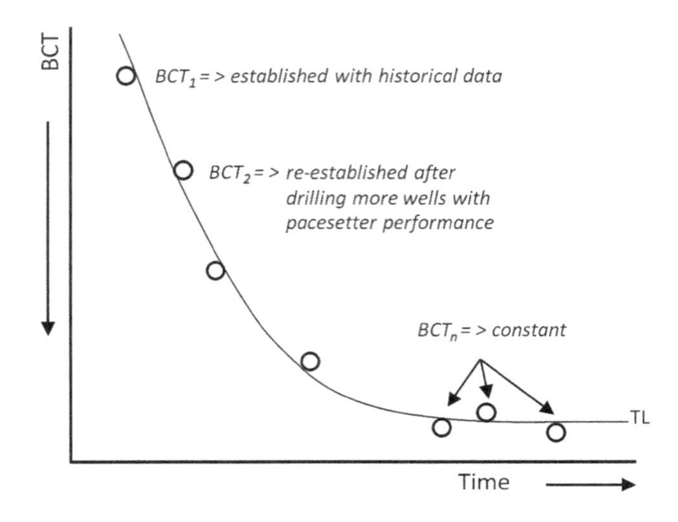

Figure 16: BCT trend as more wells are drilled (Adeleye et al., 2004)

The data analysis aiming to reduce operational duration, trouble time and inviable trouble is discussed in more details in section 5.1. Other methodologies, such as the well delivery process (Wardt 2010), present similar concepts to improve the drilling process.

4.2 ACTIVITY REPORTING

The industry commonly accepts that information from actual drilling, completion and workover operations should play an important role in the analysis of performance, comparison against plans, optimization of future operations, and operational safety. It also expects that considerable operational savings will come from the effective use of such information.

There are a number of ways companies collect this information. Some still use paper forms and fax/email them to the office, others are using Excel™ files, and the high-end users employ sophisticated software systems such as WellView™, OpenWells™ and IDS Datanet™ between rig and office either via the web or other means. All of them are filled in by hand using some sort of form. Currently, this information can be further refined with the interpretation of real-time data from mud-logging and other sources, which complement - and sometimes contradict - the human-based report.

Historically, the activity report or "morning report" was used for the office people to understand what happened in the last 24 hours at the rig site. Once they digested this report during their morning meetings the information was discarded. Little did they know that so much value could have been retrieved from this ignored information. Only now, with data management processes and clear BI objectives, is everyone appreciating the value of such historical data.

In many cases, however, companies cannot experience these gains, even when using sophisticated reporting systems and software tools, mainly because of inappropriate data management processes. For example, sometimes companies do not store detailed planning information in their databases and thus cannot compare planned vs actual. In other cases, even when planning information is kept, it is difficult to match it with actual operations, as the operations are not identified or coded uniformly, making it unfeasible for the user to perform any reasonable analysis.

The availability of good historical information opens the opportunity for the realization of the "knowledge" cycle (Figure 17) of operational improvement and cost reduction. The plan and execution sequences are intertwined processes, being the opposite sides of the same coin. However, several factors might preclude the availability, accessibility and quality of such information.

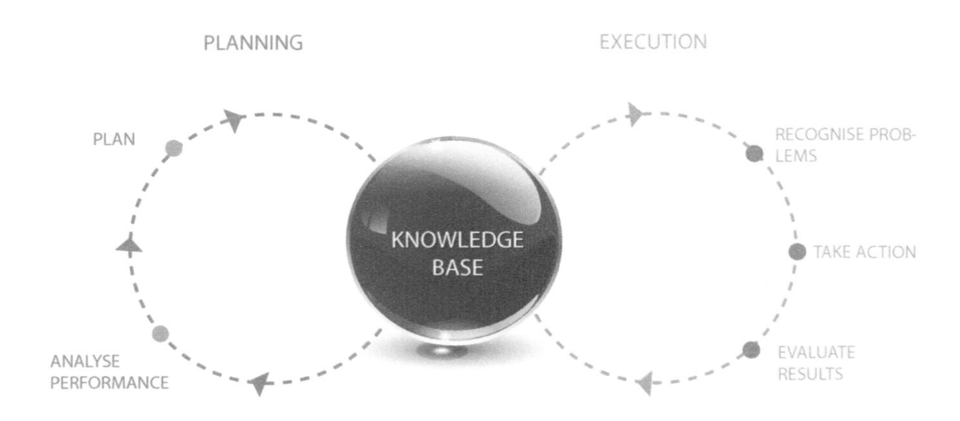

Figure 17: The "knowledge" cycle of intervention planning and execution

Nakagawa's (Nakagawa, Damski and Miura 2005) proposed solution to reducing or eliminating these negative factors employs the concept of "report on exception" planning. In that innovative article the importance of the planning sequence was clearly demonstrated as the source of the activity report. It reaches the point where the "rig daily report" would only involve filling the actual duration on the planned sequence. Any deviation from the plan is considered an abnormality to be analyzed and used as feedback for the next plan.

Drilling, completions and workovers are different types of well intervention. Each intervention contains a sequence of phases and operations and each operation contains a series of steps. Interventions, phases, operations and steps constitute different levels of activity.

Historical data analysis such as performance evaluation, benchmarking, risks, time and cost distributions and Pareto, allows learning and optimization of surface and down-hole processes including design and operational aspects. For example, from post-analysis we learn where we are now and how much it costs. From benchmark analysis, we gain an insight about where we should be, why we are not there and how we could get there. From analyzes of abnormalities, we learn about problems that occurred in the past, how we treated them and how we could avoid them in the future. With Pareto analysis, we can determine the most significant abnormalities and treat them for best cost/benefit improvement. In addition, historical data with the right pointers gives us more useful and transferable knowledge from one situation to another. Finally, by performing statistical analysis of abnormalities using operations as a starting point, we can statistically evaluate the risk of each operation and use it to mitigate the operational risks and to increase overall safety.

The main limitation for proper use of historical data is accurate collection and the lack of quality in the operational coding process for intervention activities, thus preventing the assembling of consolidated and reliable information based at the operations level (Figure 18). Such a problem is prominent in completions and workovers, although several companies still have considerable problems with drilling coding too.

Figure 18: Main barriers for the proper accessibility and use of historical data

Activity coding is very important as all statistical interpretation is based on codes. Almost every company uses some sort of coding system. Some of them are not that accurate because of:

- Lack of right granularity – this means they are not detailed enough to differentiate some aspects of the operations or, on the other hand, they can be detailed to excess making it difficult for the coder to identify the correct code;
- Inconsistent codes – as coding is (mostly) done by people interpreting the operation, they can assign different uses of codes;
- Differences in the start/end points of operations being coded;
- The human factor in coding activity – the more people coding, the more difficult it is to have the same code for the same operation;
- Misleading codes – sometimes the code does not reflect what actually is happening, for different reasons, especially relating to trouble times;
- Coding trouble codes – this is another source of misunderstanding. As a general industry parameter, it can be considered that 20-25% of the total time to drill a well is trouble time. This is an enormous amount of time and cost, which can be reduced by correctly understanding the source of the trouble and having a process in place to mitigate it in future.

Some of those points are difficult to standardize. Large companies, many assets, people from different backgrounds, lack of training, etc, are natural barriers to making a uniform, reliable, consistent set of codes. Nevertheless, this should be the goal of any company. They should establish committees to study their coding system, set standards, organize training and spread the message throughout the organization. Basically, there are no right or wrong codes; there are only adequate codes for the company and its environment and objectives. It is very important for the codes to be used in a uniform way so that later on they can be combined in statistics without hesitation. One way of addressing this problem is to set up a specialised task group charged with revising or recoding operations and applying the correct codes to operations and trouble identification.

The daily drilling activity report is the cornerstone of the planning and most of the analysis of drilling data. The following sections will describe some results that can be derived from this report. We assume that each and every operation in the field has a code and duration. Codes can be for normal and trouble times. They also have recorded the range of measured depths that they were executed at. It is important for depths to be documented as they correlate to everything else in the well such as formation, lithology, mud, bit run, etc.

4.3 PLANNING AHEAD

Forecasting does not mean avoiding risk or preventing the impact. Its aim is to recognize their probability and ensure the conditions for decisions that maintain the effects are kept to acceptable levels for the business. Forecasting is based on maintaining dynamic processes to analyze, develop, document and maintain updated scenario studies. One of the most important scenarios under consideration while developing an oil field is the planning of the duration of a well intervention (whether drilling, completion or workover).

All oil companies describe what occurs in a well intervention using daily drilling reports. As explained in sections 4.1 and 4.2, the main objective should be to transform this data from a daily drilling report into a "knowledge base", which can be used to draw the risk scenario that predicts the total duration of a new intervention. I use the term "knowledge base" here instead of "database" just to differentiate the added value the QC and coding/aggregation processes add to the "database".

This section describes how to plan an operational sequence and authorization for expenditure (AFE) for a new well, and also make provision for follow-up of the plan that will result in drilling optimization.

4.3.1 PLANNING A NEW INTERVENTION

Well planning plays a crucial role when it comes to time and cost improvements in drilling operations. Under the engineer's supervision/approval, an estimation of time can be successfully achieved using the planning tool from a group of similar wells (historic data) and a pre-defined drilling sequence (defined by the company/user). It is important that the historical wells selected to perform the time estimation are comparable in terms of phases, trajectory, location, etc.

For example, the rate of operation "running casing" can be collected per casing size from offset wells and fitted into a statistical distribution curve (for example, lognormal distribution). The parameters' mean and standard deviation of this distribution can be simulated to predict planned values. Each value can be multiplied by the total length of the planned casing string to achieve the estimated time to run this operation. Monte Carlo simulation is a well-known sampling process (see Appendix A – Statistics 101). This entire process is summarized in Figure 19.

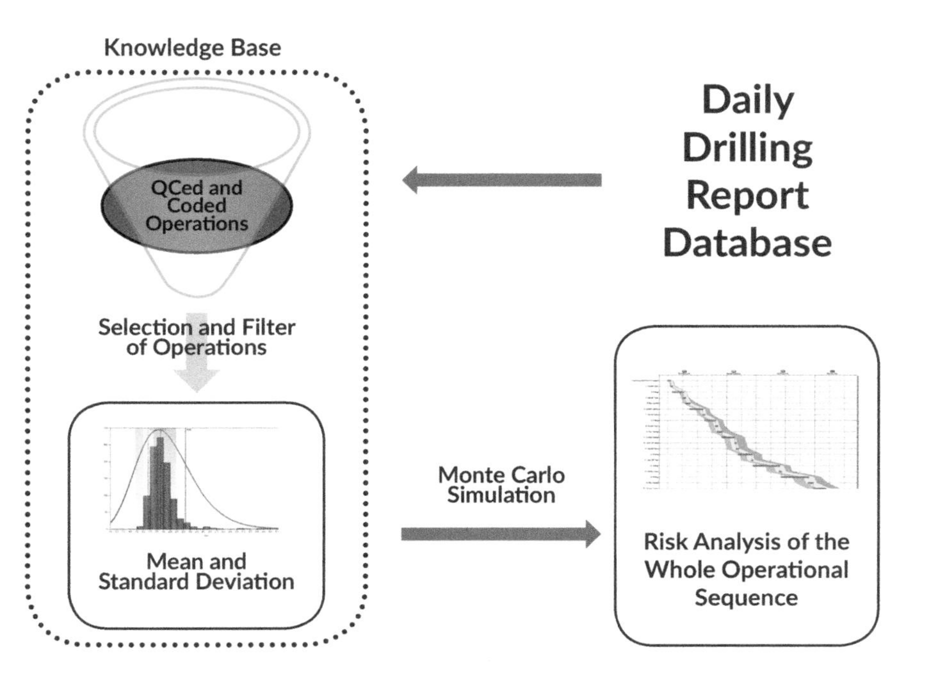

Figure 19: Process to generate a planned well sequence of operations

There are many other issues in matching historical offset wells to the new planned well. There is a need to do an accurate phase matching as not all wells use the same hole sizes, sequence of phases, etc. There is also a need for careful treatment of the historical data in a way that allows for it to be re-used appropriately in the new context of the planned well. This is valid for both normal and for trouble durations. There are many variants of such mathematical models that are beyond the scope of this book, but good software products should be able to handle most of the cases.

As a result of all this data preparation, there is a planned sequence with the uncertainty of total duration. This is achieved after performing a Monte Carlo simulation, which can produce P-values such as P10, P50 and P90, demonstrating how the uncertainty spreads as the operations accumulate. This also includes how much trouble time is expected. Some trouble time is due to unexpected issues, such as "waiting on weather" and others, are expected with more certainty such as "stuck pipe" trouble as a consequence of a particular formation. Once the engineer has a very good understanding of this spread, he/she is in position to decide for the best planned time for this well. A final result can be completed after investigation of the technical limit, learning curve, a particular P-value to be used, and other conditions. As usual, the user has the final decision on the plan, using the software as supporting tool. The planned duration and uncertainty around it would look like the Figure 20.

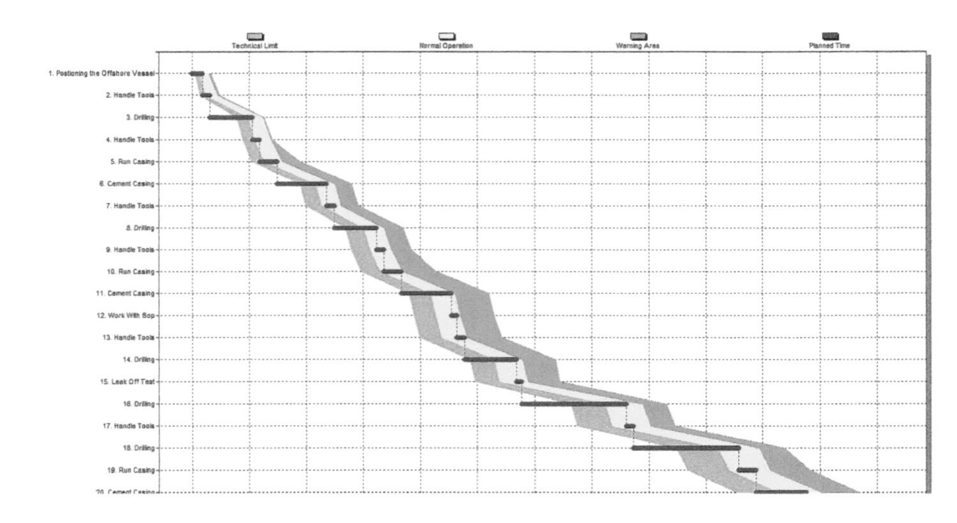

Figure 20: Operational sequence for new well based on historical offset wells and Monte Carlo simulation

This plot represents a very powerful instrument, which helps during the decision-making process, because it gives a comprehensive evaluation of the possible scenarios. The resulting outputs are in the form of distributions that include most likely values, means and percentiles that will build the basis for the planned costs. This means that the time vs depth function of the planned well will include better time, and therefore better cost estimates than are achievable through conventional deterministic methods. AFE prediction is discussed in 4.3.4.

In addition to the basic sequence, some users might need to introduce conditional situations that might happen in the process of executing it. This is to cater for some unknown situation during the planning phase. It can be implemented using an "IF ... THEN ..." structure in the sequence of planned operations. This structure uses:

```
IF <a certain condition happens> with probability X%
THEN {do those operations}
ELSE {do some other operations}
END IF
```

Although this leads to much more flexibility in the planning sequence for those difficult and expensive wells, it also introduces the more complex simulation of taking into consideration the probability of each branch and final expected duration. This type of flexibility, and others such as parallel operations using dual-derek rigs, are the features engineers can obtain through high-end software solutions.

4.3.2 FOLLOW UP THE PLAN

The drilling plan carefully projected as described in the previous section needs to be thoroughly followed up so we can learn from any actual deviation for the benefit of the next well. This is a difficult task for most daily drilling report systems, as they do not have a precise way to match planned and actuals on each operation. They are focused on "reporting" what is happening in the field, not on what "deviated" from the plan. This is the big "flaw" in the workflow that needs to be corrected. In principle, we could assume the "report" is actually already prepared in detail in the office and only corrections are to be made on the rig. This is similar to a flight plan from A to B.

It is important to know how much we deviated from plan. Even more important is to know "why" we deviated from the plan.

There might be a few reasons for that:

1. The planned operation did not happen;
2. The executed operation was not planned;
3. The planned operation was executed, but there were problems associated with it.

Case #1 is simple. If this operation was not needed, then it should be removed or be conditional (under "IF condition" in the sequence) for the next well. There is a "lesson learned" right here, but there is more subtle information in this case. Usually managers rush to see the total planned time against total executed time as a metric (or KPI) for the well follow-up. But if the operation was not executed it should not be considered as part of the comparison when the focus is on the execution. It would be unfair with operational people to state the well deviated "35% more than planned" when some planned operations were not even executed. This could be because of a bad plan, not bad execution. So, a KPI for well performance needs to consider this situation. There is a difference between a good/bad plan and good/bad execution.

Case #2 covers unplanned operations. Most of the time it should be considered as an abnormality (trouble). If the plan was carefully prepared, then any deviation is an anomaly in the process. This anomaly can emerge from the planner, who forgot or missed this operation, or because an additional operation was required due to an unforeseen situation or bad execution somewhere in the sequence. Regardless of the cause, it needs to be registered and understood, so that the plan for the next well takes this situation into consideration.

Case #3 is the classical example of trouble time during the execution of a planned operation. In this case the executed time is recorded separately by normal and trouble time (if any) and it is possible to effectively measure the deviation from the plan. It might be important to associate the type of trouble involved in this particular operation, so the revision in post-analysis can judge if the trouble can happen in the same operation in the future or not. The total normal time can give an indication of performance if compared to the history of this operation under similar conditions.

All these three cases demonstrate how important it is to follow up executed operation on top of the plan. There are many lessons to be learned from this comparison and feedback for the next well. It also allows more precise measurement of how efficient the drilling campaign is in detail.

But the efficiency of operations in the well does not finish within the well itself; it needs to be compared with other wells in a similar situation. It is rational to say "operation A" took 10 hours with 3 hours of trouble time, and it was planned for 5 hours, therefore it was a catastrophic execution.

If we compare this with offset wells for the same operation, the statistical results might show a different perspective. If the same operation has P10, P50, P90 with 3, 7, 20 hours respectively, then the result was not that bad and the plan was too "tight". But we cannot restrict this analysis to a focus on a single operation. We need to see the whole picture and also see how other operations went.

Figure 21 shows a sequence of operations, with the "box plot" representing the statistical distribution of each operation (from a historical perspective) and the planned and executed durations.

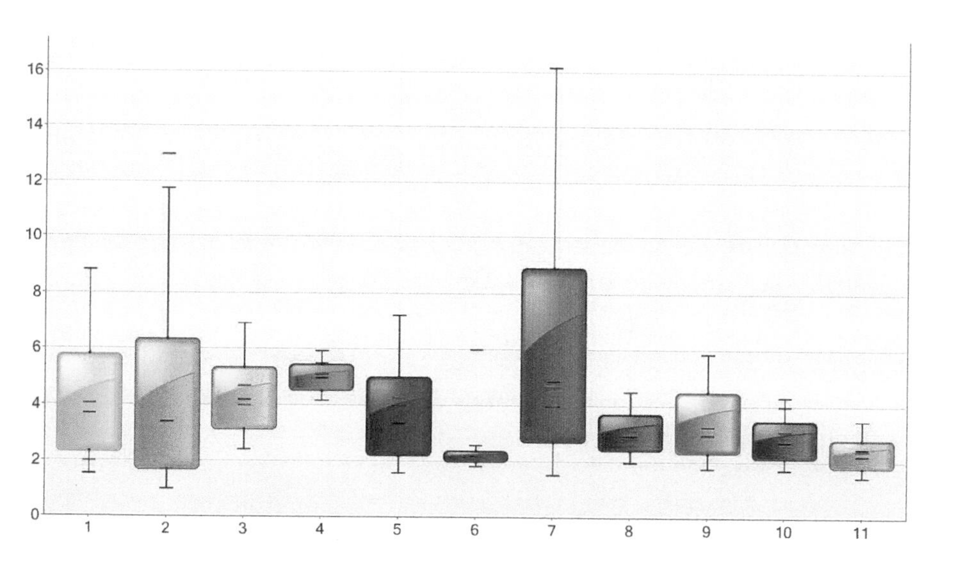

Figure 21: Sequence of operations against duration in Y-axis – box plots show historical time distribution and flat point the executed duration of each operation

With this information it is possible for the engineer to draw better conclusions about the well performance, not only against the plan, but in historical perspective. For example, operation "1" went exceptionally well, performing below most of the offset wells, while operation "2" had the opposite result. Percentage wise, the loss in operation 2 was much higher than operation 1. Looking into the specifics of each operation, the drilling team can keep doing what is working well and figure out ways to improve the bad results. It is simple and effective, but a good data management system and procedures need to be in place to achieve this level of understanding.

4.3.3 DRILLING OPTIMIZATION

In reality "drilling optimization" needs two consecutive steps:

1. Have the process of each operation under control;
2. Reduce the time to perform an operation.

Step 1 is usually overlooked as most people are focused on reducing time. It deals with the uncertainty the operation has been executed. Assume that operations 2 and 5 in the Figure 21 have around the same P50, but P10 and P90 are much more dispersed in operation 2. The difference between P10 and P90 indicates the control over the process of executing this operation. The further they are apart, then the further the operation is out of control. That means people in the field can execute an operation in 8 or 16 hours! How come the same operation performed in two similar circumstances can take twice as long depending on slightly different conditions? We would need to go into the details of each data point in operation 2 and discuss technically what is going on to understand this difference.

Figure 22 shows a typical example of a drilling sequence. It is easy to spot that it is more important to work on operation numbers 13, 26 and 39 first, as the P50 and variance are much higher than in other operations. If engineers had not seen the operations in such a format, they would not be able to work on them. There is not much time to ponder, so we need to focus on what matters most.

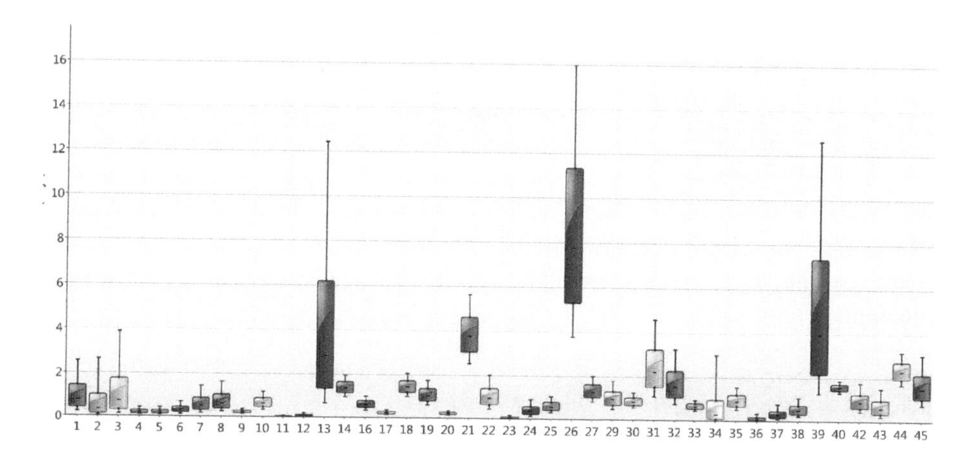

Figure 22: More complex sequence of operations

But the important point is in reducing the P10 to P90 spread, having more control over the process, and therefore the ability to forecast the next well more precisely. The predictability of future wells is as important as ensuring the operation is running at optimal performance. It gives more certainty for AFE, rig management, logistics, etc. The whole process works better if the total duration of each operation; hence the whole well, can be predicted more precisely. The same concept also works for the entire set of wells. It is possible to combine several similar wells and create a box plot of distribution of their total time. In this case, a campaign of dozens of wells can use a similar approach, as was demonstrated by (Hougaz, et al. 2012) in a sequence of wells as shown in Figure 23.

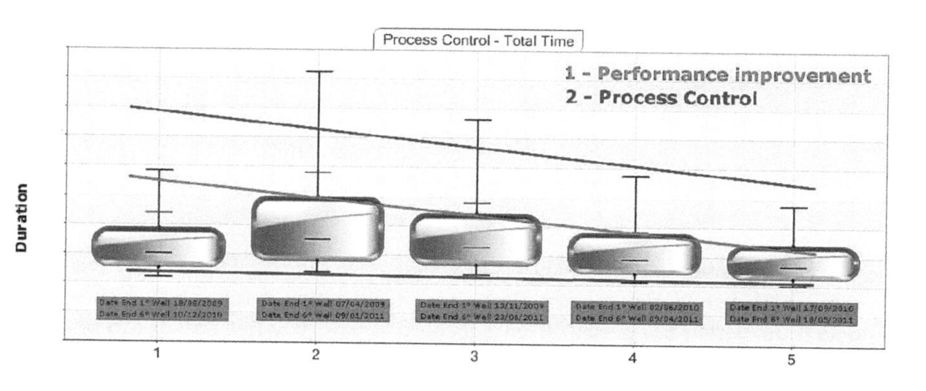

Figure 23: Sequence of wells drilled

Figure 23 shows a set of executed wells with their total time distribution represented by a box plot from P10 to P90. Each case in the sequence 1, 2, 3, 4 and 5 are sets of wells. The blue lines are linear "best fit" for P10 and P90, respectively. The convergence of those lines clearly shows the improved control, as the spread between P10 and P90 is reducing over time. The red line is the linear "best fit" for P50.

The second step is the optimization of the process. Once each operation is under control, we need to reduce its duration. This is measured by monitoring the P50 of the operation over time and also observing the learning curve. Figure 24 shows the statistical duration (in the form of a box plot) of a given operation in a sequence of five wells, ordered by spud date. A simple trend line shows a decrease of P50, so the actual duration (or rate - which is inverse) is getting better.

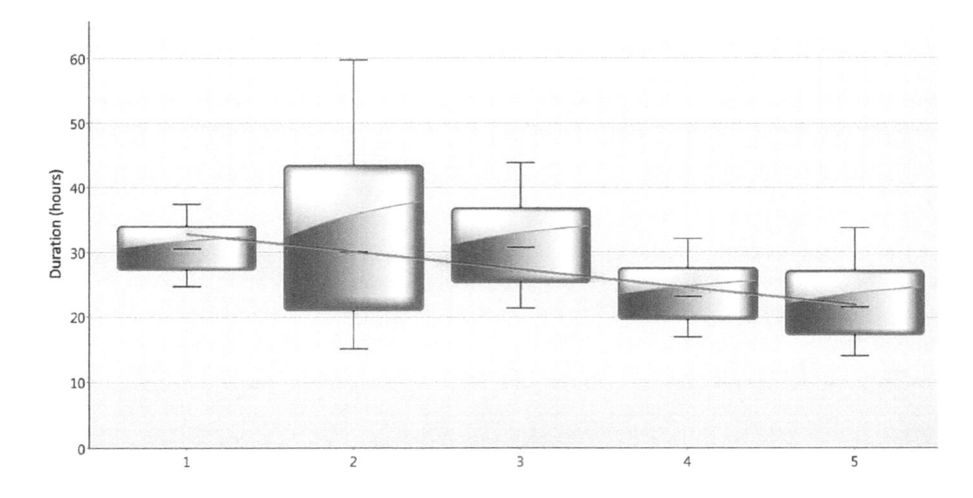

Figure 24: A particular operation executed several times in each of five wells, ordered by spud date

The first step is to technically analyze the cases above the trend line of this set. If the trend line is already declining, then there is a need to look in the learning curve plot to understand where the optimization is located in time. It is not possible to optimize forever, but in a particular timeframe. To reduce the duration even further there is a need to change the underlying technology, which leads to another cycle of process improvement.

4.3.4 GENERATING STATISTICAL AFE

Martin Hanschitz (1997) did great work explaining how to generate AFE. It is common knowledge that at least 30% of the total cost of the AFE is time-dependent, especially in the market of expensive rigs. If such a big component of the AFE is time-related, one cannot generate just a deterministic AFE. Financial managers need to understand the uncertainty around the planned AFE, for the best and worst case scenarios. This is relatively simple to do if you have the right software tools. Suppose your planned time vs depth sequence is simulated and has a total distribution as shown in Figure 25. The chosen planned time is somewhere around P50 as the most likely duration for this well. To get the statistical AFE, it is a matter of using desired lower and upper boundaries, such as P40 and P70, and recalculating the time-dependent items in the AFE by those new durations. This mechanism may, or may not, incorporate trouble time expected for each phase or well, in the same way that we may, or may not, incorporate trouble time in the TxD curve.

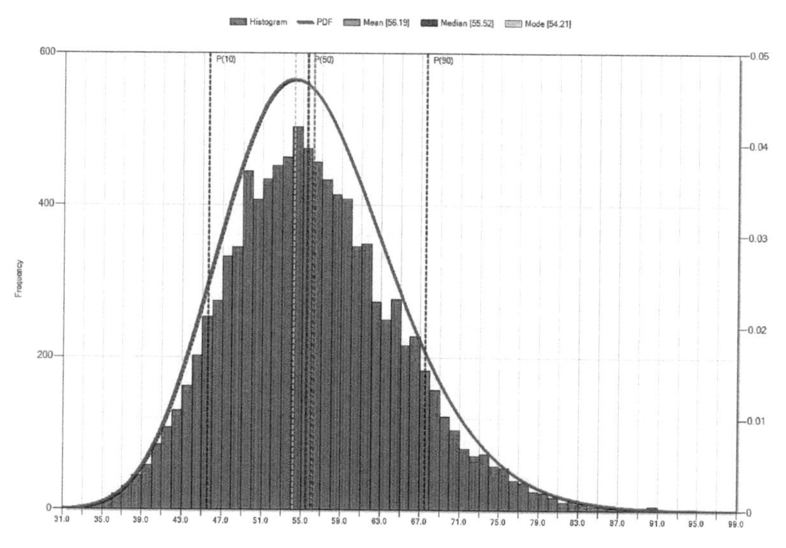

Figure 25: Well total time distribution

In real cases, this process can be a bit more complex. Some items can be dependent on a particular operation, or a specific phase, rather than the total time of the well, but the same mechanism is applied. Jim Murtha and others have extensively published models to generate statistical AFE, such as in the paper SPE 26339 (Peterson 1993).

In any case the final desirable summary of the statistical AFE would look like the table in Figure 26.

PHASE NO.	WITH TROUBLE			WITHOUT TROUBLE		
	Mean Cost	P5 Cost	P95 Cost	Mean Cost	P5 Cost	P95 Cost
1	$320,968	$207,336	$517,068	$290,864	$192,496	$485,268
2	$464,515	$357,600	$616,967	$347,701	$318,307	$367,805
3	$1,017,296	$957,895	$1,079,049	$919,267	$884,535	$937,283
4	$1,546,904	$1,017,488	$2,305,992	$851,960	$714,400	$923,552
Total	$3,349,683	**$2,540,319**	**$4,516,076**	$2,409,792	**$2,109,738**	**$2,713,906**

Note: Only Mean Costs sum to total. Other values are approximated.

Figure 26: Summary of statistical AFE

Any good AFE generator product can describe all cost items in terms of financial centers such as administration, logistics, third party services, etc. This is normal for accountants, but more interesting is to re-structure this AFE and show it in terms of operations and phases. In this case the user can see the cost at operation level that can sum up to phase and well level.

Almost any AFE system can allow for daily cost follow-up as per field tickets reported, and also some support to integrate with the real costs. This gives a portrait of "estimated" and "real" costs. The estimated cost, based on information from the rig, is just to keep everyone informed of the running costs. This is appealing for the personnel working on daily well operation. The "real" cost is related to the actual accountancy values, which end up with precise costs, a long time after the well has finished. This is important for accountants but has almost no value to engineers in their daily operations. Engineers are interested in comparing statistical planned AFE against daily estimated cost during operations' follow-up.

Unlike the operational sequence of the well, the planned AFE is more difficult to generate in terms of statistical forecast from previous actual AFEs. Fluctuation of the currency exchange, a different third party contract basis, day rig costs, etc, are dependent on market forces and other parameters that are beyond the company's management. Those parameters are difficult to forecast and to associate with uncertainty. It therefore seems more reliable to use the time variance of operational offset data to forecast the new cost than to use the dollar (or any other currency) as the basis for the statistical model. The unit cost needs to be given by the user as input for the specific situation of the new well.

4.4 MAP INTELLIGENCE

It is common knowledge that mapping plays an important role in oil & gas. There are many complex and comprehensive solutions available on the market to cover most of the functionality needed in this regard. There are a few BI solutions among them catering for some drilling needs. There are at least three examples:

- Plots at survey level;
- Geo-located plots and trends;
- Geo-located documents.

At survey level there is a need to know where particular problems can occur against a particular direction or an inclination in a given region. Everyone is familiar with 3D projection against formations, etc, but there might be more than that. Figure 27 shows a simple example on a 2D projection of a well's survey and displays with different colors where particular problems occur at the survey position. Locating the trouble type, such as stuck pipe, borehole stability, drag, hole cleaning, etc, and considering inclination and azimuth against measured depth can help engineers to identify areas and trends in the field and mitigate them in future wells.

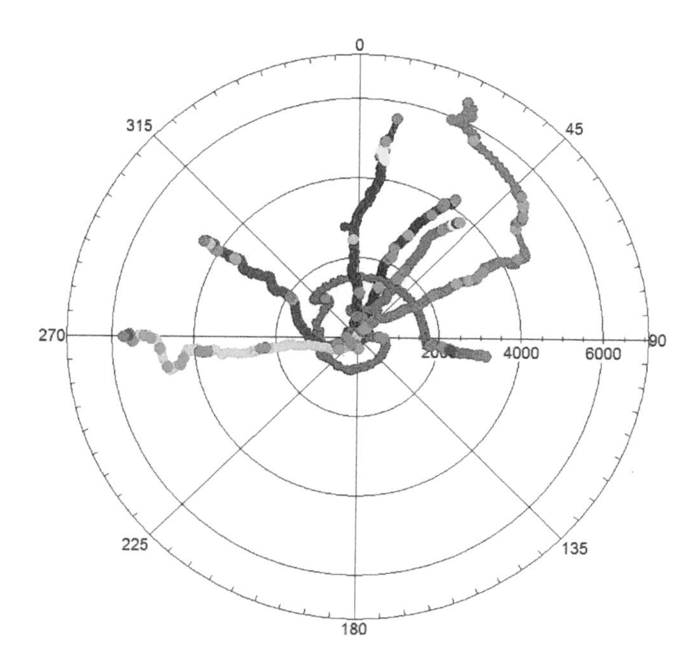

Figure 27: Example of a 2D projection azimuth vs MD (m) of some wells and where different types of trouble occurred

This can also be coupled with electronic log information. Log analysis is a comprehensive field by itself and it is beyond the scope of this book. Electronic logs are used for specific purposes, such as assisting drillability and understanding geomechanical issues, etc. But there are other aspects in the use of logs concerning cross-referencing with other data as a result of better data management. For drilling engineers it is important to compare offset wells regarding their formation, lithology, mud properties (weight, types), survey data (inclination, azimuth, DLS), any electronic log data, and events (such as troubles, tests, etc) that occurred in each well. Figure 28 shows a typical plot with three wells side by side.

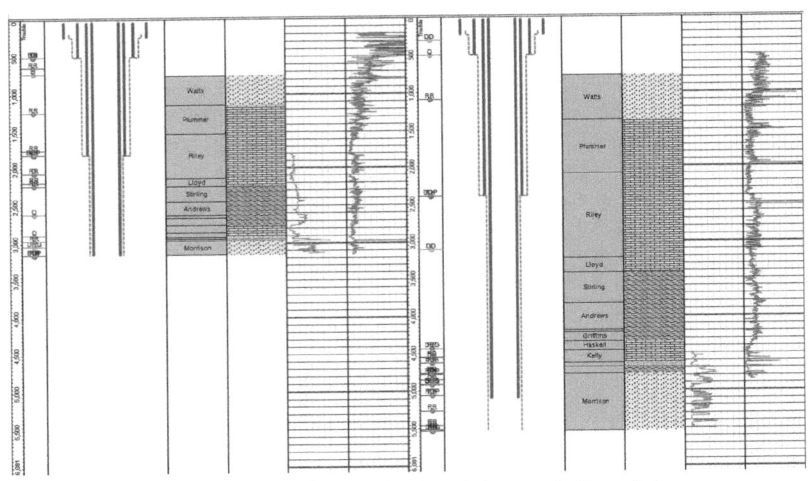

Figure 28: Correlating electronic log and other data

In this scenario it is possible to identify trends and correlate them to the next planned well, choosing better casing shoes, bit change, mitigating geomechanical problems, etc.

The second level of map visualization is related to data aggregation at well location. This is particularly important for offshore fields where there is no landmark and any underlying map shows only blue water and a grid. At the well's location users should be able to plot any data related to time (such as total, phase, operational durations), trouble time, different types of status and any other useful information, such as shown in Figure 29. This goes beyond drilling, as it can show production levels, running costs, etc, not only from wells, but rigs, floating production, storage and offloading (FPSOs), production facilities, and any others entity with a specific latitude and longitude position.

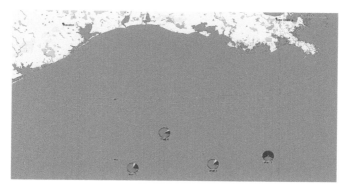

Figure 29: Example of map with added aggregated information

The third level is related to unstructured data such as documents, images and multi-media files. Everyone is familiar with document management systems, i.e. how to store and retrieve data based on keywords. But engineers might need to retrieve documents with geographic tags where well name, field name, formation name, rig name, etc, are not a string to search but a location on the map that gives them meaning. You can do this by attributing a geographic reference (such as latitude and longitude) value to a name or a file. In such a case the engineer would go to the map area of interest and search for all documents available for this spatial region. The map should show all documents (reports, spreadsheets, presentations, PDF files, etc) also multi-media files (videos, photos, plots, etc), which contain geo-location tags related to that region. Further filter conditions would prune the results until desirable documents are found. An example of such a map is shown in Figure 30. It is worthwhile mentioning that the search starts with the geographical region and not by retrieving a bunch of documents from the document management system and trying to figure out something about them.

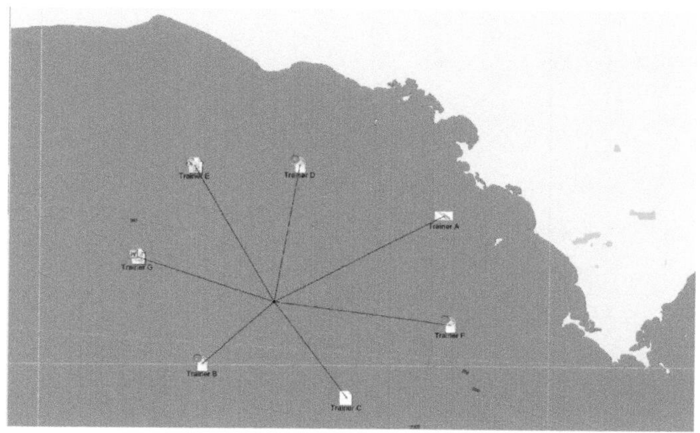

Figure 30: Example of map with cluster of documents

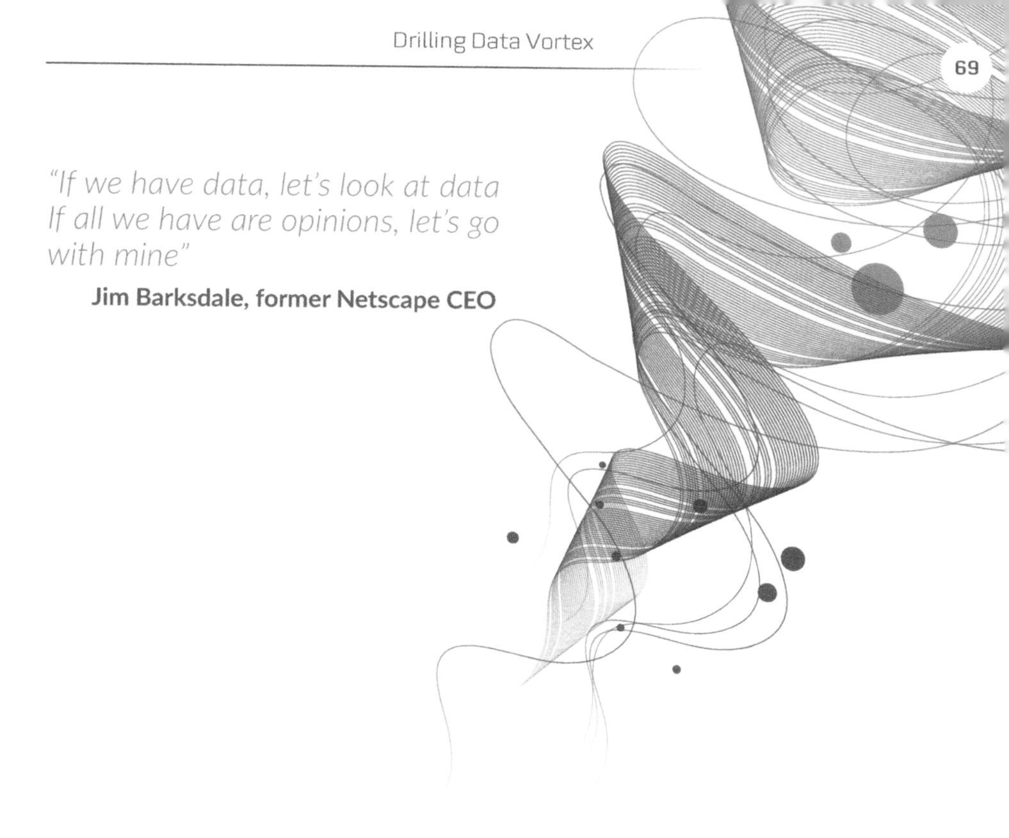

*"If we have data, let's look at data
If all we have are opinions, let's go
with mine"*

Jim Barksdale, former Netscape CEO

5. DRILLING DATA ANALYTICS

This section describes in more detail some of the most important analysis that can be performed using drilling data. Some of the KPIs and benchmarking results emerge from direct comparison between products and service providers. It is important to observe that the company performing such comparison is not related to the company providing the goods and services, in order to avoid conflicts of interest that might bias the results. That's the same reason the "auditor" needs to be independent from the "accountant", otherwise it will defeat the purpose of independent measurer.

With analytics we are closing the Deming's loop P-D-C-A (see Figure 13), in the "C" for checking the results. This feeds into "A" for action based on this resulting data. Data needs to flow from one stage to another. If it stays stagnated in one place it has no value. The faster we can loop through PDCA the more value we can add to the business.

5.1 TIME AND TROUBLE ANALYSIS

To reduce the gap between the current well duration and the ideal target (technical limit), a comprehensive analysis of the activities developed during drilling operations has to be made to identify improvement areas or those areas that the company has effectively performed. You do this by having a consistent operational coding system and the ability to reduce NPT and invisible lost time, as discussed in section 4.1. The first step would be to identify the most important areas that can be improved. The first direct tool to use is learning curve plots for all levels starting at total time of the wells, down to major operations and phases, and finally at operational level. This points us to areas of deficiency, but not which ones are more important to start working on. The learning curve analysis is described in detail in section 5.2.

Another tool can be used during the post-analysis to make comparison of multiple wells in different scenarios. In Figure 31 a time breakdown of the major operations is observed in a set of wells, where it is possible to identify and focus on those important areas that need further enhancements. It can be easy to observe the operations that comprise most of the total time, and then the engineer should analyze these operations in order to examine how the time was spent and detect possible problems/ improvements.

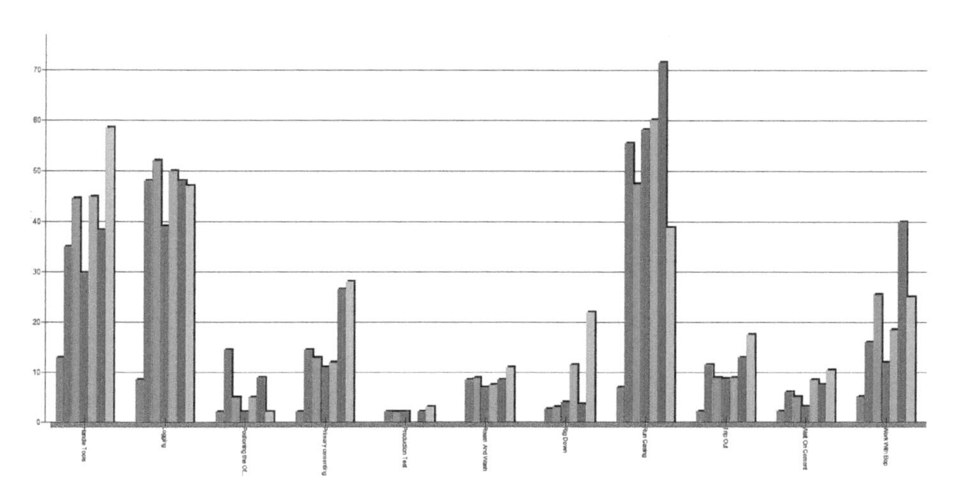

Figure 31: Post-analysis plot of the major operations duration groups by several wells

Once the important operations are selected, the next step is to plot a histogram with all cases in the historical offset wells, as shown in Figure 31(a). The lognormal curve helps to see the mean value and the spread, so the user has a perspective of time variance. In this case there is a need to investigate specifically the worst cases to identify the reason for long durations (trouble, low performance, etc) and try to mitigate the issues in the future. The learning curve (LC) lines in Figure 32(b) give the duration behavior over time and draw some expected future time and its variance for this operation.

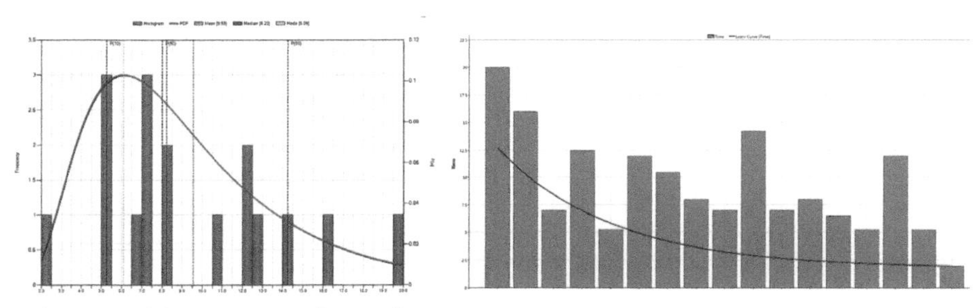

Figure 32: (a) Histogram of a single operation across multiple wells

Figure 32: (b) learning curve

In an attempt to identify and pinpoint the causes of the problems in a well's performance, the trouble time breakdown can be used to determine the magnitude and repetitiveness of the main issues, to check negative trends causing negative learning and to identify significant opportunities for improvement. The first step is to identify the most common trouble across multiple wells. A simple plot with total trouble duration and number of cases helps engineers detect the most relevant problems in the field, as shown in Figure 33.

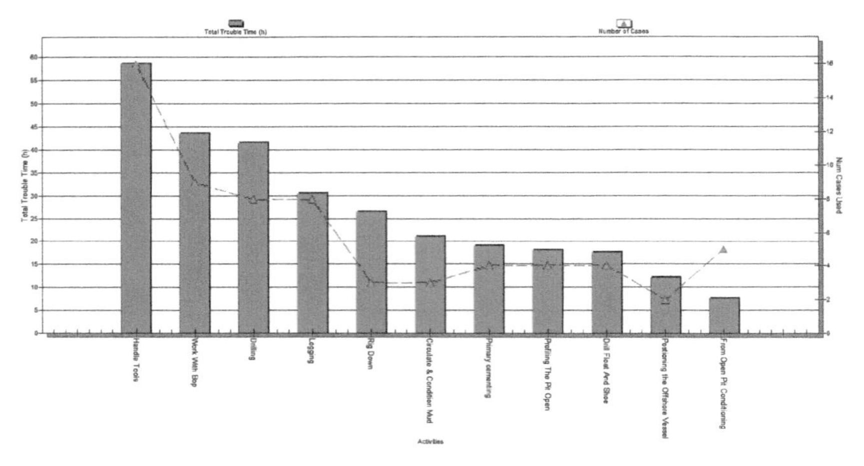

Figure 33: List of most important trouble times in descending order

This information can be complemented by the occurrences of each trouble time per trouble type in each well under consideration, as shown in Figure 34.

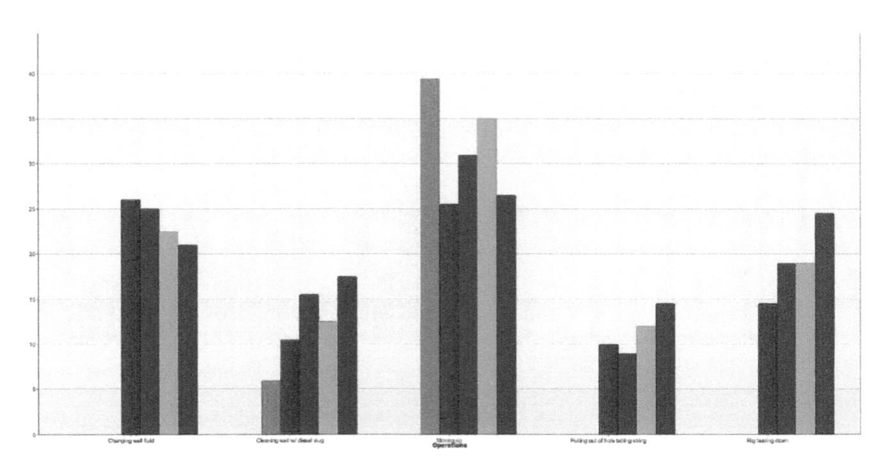

Figure 34: Trouble occurrence per trouble type for multiple wells

Once the main problems have been detected, further investigations must be performed to pinpoint their causes and plan different ways to mitigate them. For example, time vs depth graphs can be used to relate the problems to the depths and observe the zones where they have occurred. A multiple well time vs depth graph can be created, as shown in Figure 35, to compare trouble type and frequency between different wells.

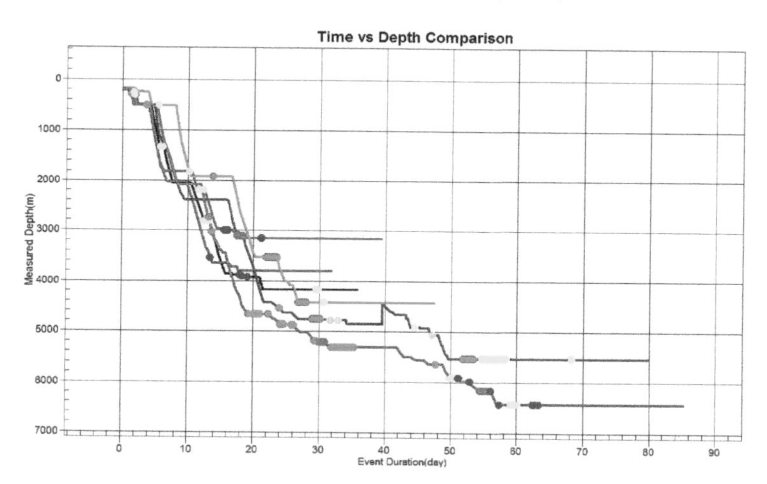

Figure 35: Multiple time vs depth plots showing selected trouble types

Most common troubles fall into the following classes:

- Waiting on weather/environment – there is not much that can be done apart from understanding the seasonality of this trouble;
- Waiting on resources – most of the time this is a logistics problem and parts of this can be solved using better well planning. An accurate drilling plan and an effective daily follow up of this plan can give good information to the logistics department on when operations will happen, allowing them to plan accordingly;
- Problems related to geology and hydraulics – these can be correlated with plots between mud-logs, bit data, geological data, trouble type etc, against depth. This feature is described in section 5.3 about bit performance;
- Rig-related problems – downtime, maintenance, etc – there is not much that an operator can do apart from reducing the cost of the well using contractual terms;
- Problems related to trajectory – these can be linked to the inclination/azimuth, as shown in Figure 27 and Figure 28. In that plot type, several wells can be represented and it is therefore possible to detect if there is any relationship between a troublesome operation and azimuth. If there are any trends, it would be possible to recognize certain areas in the field that are prone to experiencing particular problems.

5.2 LEARNING CURVE

Learning curve analysis is a powerful tool that is used to quantify a company's performance on a drilling campaign, based on a set of comparable well data. The theory of applying learning curve analysis to drilling times was initially discussed by Brett and Millheim (1986) as a tool to judge the drilling performance of a company drilling consecutive wells in a given area. The learning theory postulates that the time required to drill a sequence of wells in an area declines exponentially, which can be defined mathematically by the formula:

$$t_n = C_1 e^{C_2(1-n)} + C_3$$

Where tn is the total time to drill the nth well, and C1, C2 and C3 are defined as the learning potential, learning rate and operational limit. C1 implies a readiness, ability to be prepared, difficulty of the area or technology. C2 relates to how fast the organization learns and C3 is the level of performance for any company or contractor. The meaning of these parameters is shown schematically in Figure 36.

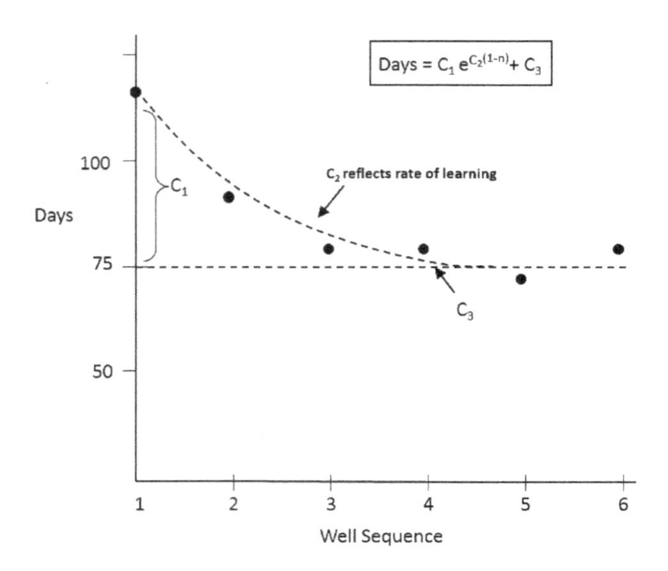

Figure 36: Typical new project learning curve (Millheim, Maidla and Kravis 1998)

An indication of the company's performance is based on these constants. C1, C2 and C3 can therefore be improved as a result of better initial well planning, more experienced staff, and/or better equipment. The key to this improvement is based on the information and knowledge obtained from previous experiences.

Figure 37 illustrates an example of a group of wells ordered by spud date using the learning curve concept, with total time normalized by total depth (where applicable).

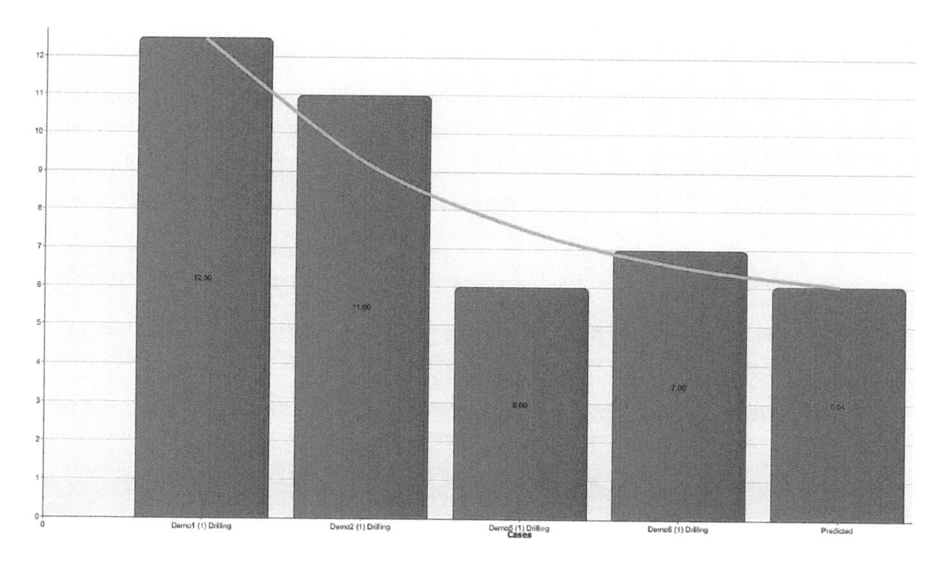

Figure 37: Learning curve showing organizational performance improvement

From this example, it was established that the values of C1, C2 and C3 were 7.1 days, 0.6 and 5.4 days, respectively. The next predicted time for this operation is 6.04 days.

On each new well drilled we are therefore trying to achieve the maximum drilling learning rate (C2) to reduce the gap between C3 and current drilling performance at the maximum speed possible. Iyoho et al. (2005) revealed that organizations with C2 between 0.45 and 0.8 show a good performance and that a C2 greater than 0.8 indicates an excellent performance. From the results obtained from this example, it is observed that the learning rate (C2) needs to be improved and therefore the organization must learn and capture experience and technology in such a way that it can be rapidly transferred to other operational personnel (Brett and Millheim 1986).

Learning curve at total time level only shows company-level improvements. Even if there is a good learning curve at total time, there is room for improvement in sub-parts of the well. Additional performance analysis can be done decomposing the duration of the major operations, which enables visualizing how the time was distributed across the different drilling activities. Figure 38 displays the time spent on different well operations. In this plot, the black line shows good learning at total time level, but the red line shows an un-learning taking into consideration only drilling related durations, which are shown as red areas.

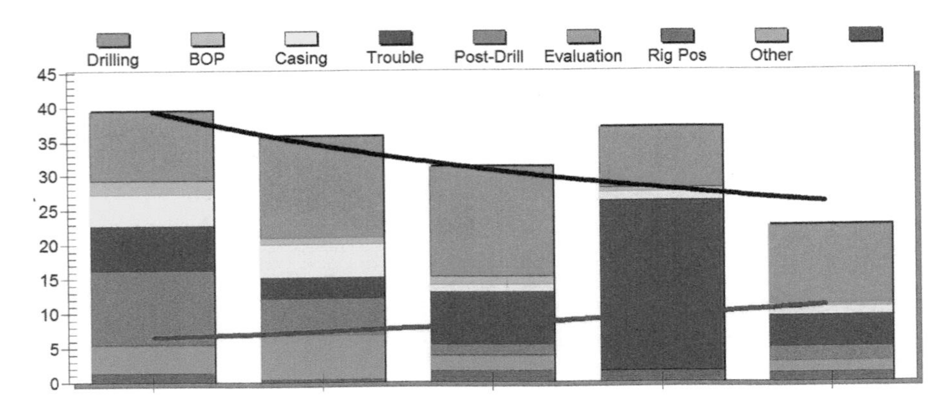

Figure 38: Learning curve plot showing the time distribution of the major operations

All in all, the general learning curve allows for the evaluation and recording of the corporation's progress over time (wells drilled). On the other hand, by breaking down the major operations and/or drilling phases, particular areas of improvement can be identified and closely monitored. The overall learning curve does not give details of the individual performance of the drilling operations. The engineer should therefore analyze the learning achieved in particular areas and evaluate what can be done to optimize to improve the overall performance of the company.

5.3 BIT PERFORMANCE

Bit performance can be interpreted in a general or in a detailed manner depending on the data available. The bits used in drilling a well have a major impact on the efficiency of the drilling phase and the costs involved. Reviewing the efficiency of the bits used in similar wells provides a better basis for selecting the new bits for the next well. When analyzing the overall bit performance it is important to take into account bits that were used for a similar formation and duration.

ROP represents a good bit performance indicative because it will show how fast the formations were drilled. In this analysis it is important to consider different bit attributes (distance drilled, formations, weight on bit (WOB), bit diameter, hole size, etc) against the ROP. Figure 39 shows the ROP of multiple 12 ¼" bits used in similar wells, ordered by ROP. There is a significant difference between their mean value and this is even worse between the highest and lowest ROP values. But, unfortunately, it does not reveal the complete picture as distance drilled and other parameters influence this measurement. On the other hand, this is better than taking decisions based on specifications from the vendors about bit performance as it depicts real performance in the field.

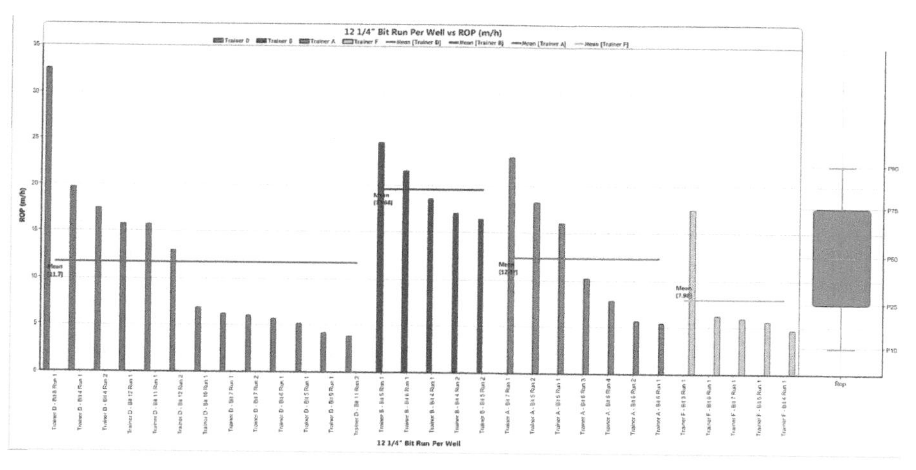

Figure 39: ROP analysis of 12 ¼" bit runs used in similar wells with mean value per well and overall statistical distribution

The drill string trips and tool handle information are also required to give a general idea of the performance of the bit. For example, if a tri-cone bit is used to drill a formation using a high ROP, it will wear more quickly, requiring another trip. On the other hand, if a polycrystalline diamond compact (PDC) bit is used a probably lower ROP will be obtained, but it will be possible to drill deeper without tripping.

To understand some aspects of bit performance we would need to see side by side the formation, lithology, survey aspects (like inclination), mud type and weight, electronic logs, among others, as shown in Figure 40. All this information gives suitable data about the bit performance under the surrounding conditions. With this, the user might remove some bit records from the Figure 38 and have a much clearer comparison between bit types and bit suppliers.

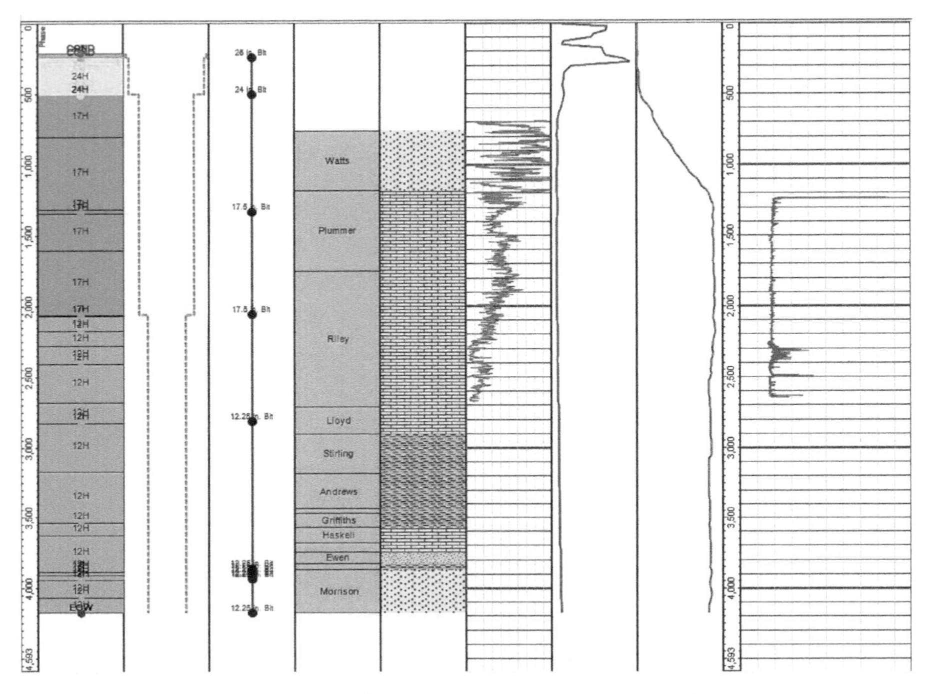

Figure 40: Evaluation of the bit performance based on formations, lithology, survey, e-logs and phases

Another side consideration in analyzing bit performance is to correlate the bit run against some reported problems at certain depths. The trouble can link formation-related problems, such as loss of circulation. The trouble events, showing the depth where the problem occurred, can be compared against the mud (weight and type) that have been used as well as the caliper log, as shown in Figure 41.

All this information is part of the overall bit performance study and it is at the fingertips of most engineers. This is the power of data management; delivering the means to compare performance and also to understand the causes underlying lower performance cases.

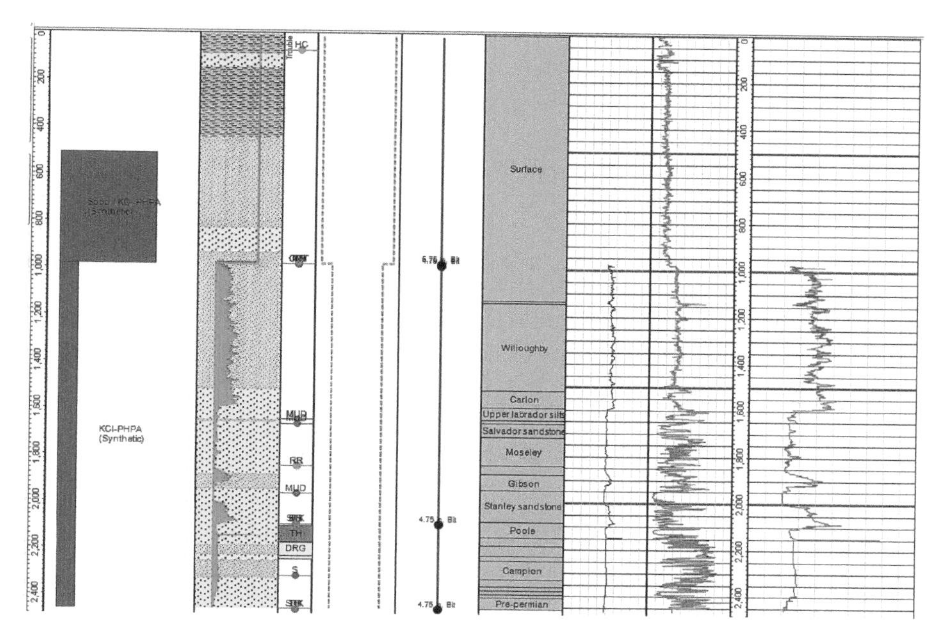

Figure 41: Plot of lithology, bit runs, mud type and caliper log, and trouble types

5.4 OPERATIONAL TECHNICAL LIMIT

Many engineers have different interpretations of the technical limit for drilling operations. They can be summarized into three aspects:

1. **Theoretical** - relates to the physical limit to someone (or something) performing a job, based on first principles (physics and engineering processes);
2. **Practical** - relates to best performance already achieved under similar circumstances;
3. **Managerial** - relates to people, providing training, motivation and other resources to determine the best that people can achieve.

As this book is concerned with data management, aspects 1 and 3 are more elusive, although they have merits on their own.

The "practical" approach simply relates to measuring the best performance in each operation for given offset wells. The final composition of all best operations is, as already noted, called best composite time or BCT.

The BCT represents an ideal case scenario, which serves as a yardstick to determine the efficiencies of operations. To obtain it, a group of comparable wells are selected and the data is broken down by activity based on the drilling sequence (for example, rig up, cementing, running casing, etc). Then, the best-ever performance recorded for each activity is identified and added up to obtain the BCT, which can be considered as the pacesetter well. This best performance is the total duration or rate for the depth-dependent operations. The resulting total time estimate is considered to be the perfect well – the best achievable with current technology and operational practices (Adeleye, et al. 2004). On the other hand, the engineer has to make sure that the calculated BCT is representative of the wells selected, hole sections, wellbore configurations, depth drilled, geological characteristics, etc. Some fine adjustment might be needed for specific conditions.

The time vs depth plot in Figure 42 shows the offset wells and the BCT (red line) obtained in this process. The BCT allows for a comparison of the performance of each well against the "perfect well" and also as a reference point when planning for the next well.

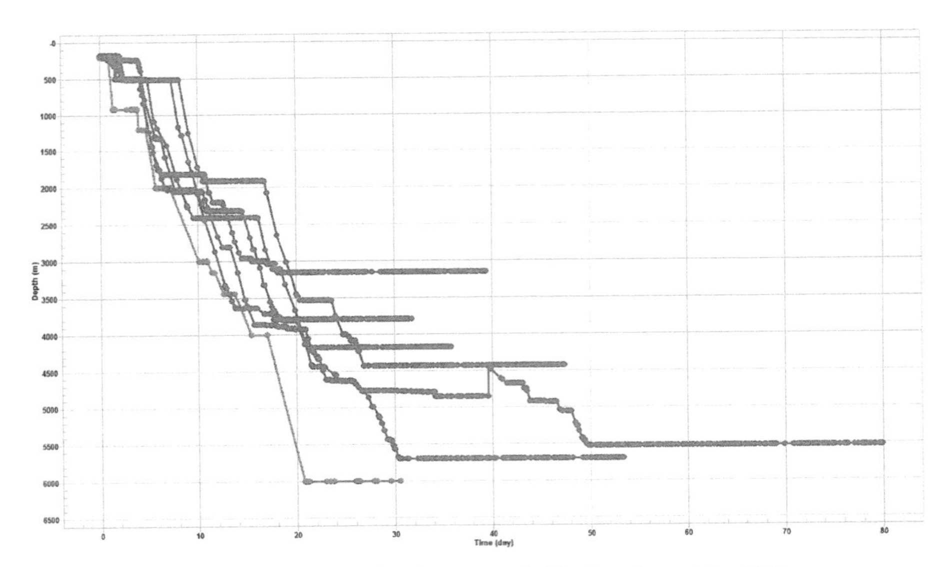

Figure 42: Time vs depth curves of offset wells and the BCT

The best way to estimate the technical limit is following up BCT through time. BCT is a moving target, which gets re-established as data from an increasing number of wells become available. During a well campaign it is worth observing how the BCT evolves as new wells are drilled. It should follow an asymptotic learning curve; hence the "technical limit" within this environment, as previously shown in Figure 16.

5.5 BENCHMARKING

Benchmarking is essentially the process of comparing a specific metric against a population of those measurements. The field is so vast that it is difficult to define. Previous sections of this book already mentioned some benchmarking. After all, those KPIs are a means for companies today to relentlessly stay competitive in the marketplace.

In the drilling process these measures are typically applied to time and cost and they can span from internal to external benchmarking. An internal benchmark is related to a company's own wells, rigs and service providers. An external benchmark is the result of comparing a company's internal data with an external source. The comparison data can be used from other companies or from industry databases such as Rushmore Reviews (www.rushmorereviews.com) and Dodson Data Systems (now IHS – www.dodsondatasystems.com). It is important to have in mind the basic steps to follow in performing benchmarking:

1. Decide what to benchmark;
2. Understand where you are;
3. Collect benchmarking data;
4. Interpret the findings;
5. Create an action plan to rectify bad performance and adopt best practices.

The "Decide what to benchmark" seems to be a simple task, but it can be deceiving. In drilling activities people usually perform metrics such as ROP (m/h), cost per meter drilled ($/m), NPT as a percentage of total time, etc. They are valid metrics, but we need to look at this carefully. For example, it is understandable for one metric to maximize ROP, whilst another one is designed to see the final result (quality) of the wellbore. Sometimes drilling too fast may cause bad wellbore finishing, which can result in more damage than good when looking at the entire life of the well. The "cost per meter drilled" can also be confusing. The cost associated with a particular currency is subject to exchange fluctuations and inflation.

Comparing "dollar" figures in a span of 5 years can produce very different scenarios. The raw number needs to be adjusted in this period to be compared. The same applies to major cost fluctuations resulting from market forces, i.e. the rig-day price also changes over time, place and types of contract. This is not to mention that companies, and even a business unit, may account for costs in different ways, for example, what and where the costs are distributed. The number is easy to produce, but its real meaning requires deeper thinking.

If users follow some of the data management best practices shown in this book, the data collection can be straightforward. Once the KPI metrics are established then we can start collecting "benchmarking data", which is located within the company or elsewhere. In drilling we can basically benchmark at these levels:

1. Total drilling intervention;
2. Parts of drilling intervention, by phases or sets of operations;
3. At individual operation;
4. Sub-parts of one operation.

Benchmarking at intervention level lump sums the whole intervention by either time or cost. Basic parameters, such as meters/day, $/meter, $/day, dry hole days per 1000 ft, etc, give an approximation for a set of similar drilling interventions - see (Rushmore 2011) for comparisons of a large number of wells.

But accurate comparisons go beyond those rates, as there is a need to normalize wells. The normalization process takes into consideration mechanical aspects of the well such as drilled depth, water depth, maximum well inclination, thickness of a particular formation, etc. For example, it can cater for sub-parts of the total time that are dependent on drilled depth, excluding those that are not. Usually, this is done by choosing a reference depth or drilled meterage, and normalizing all wells according to this reference point. In this case, all durations are comparable and wells can be benchmarked by the amount of time used, as shown in Figure 43. In this figure, it has chosen the "total drilling + flat time" duration of each well and plotted against its total drilled depth. Each well duration was then normalized to a given arbitrary drilled depth of 3000 meters and plotted as a red dot. Interesting to notice in this plot is that the wells drilled between 3500 and 4000 meters are the "best performers" (minimal normalized duration) and the next best ones are widespread in their depth range.

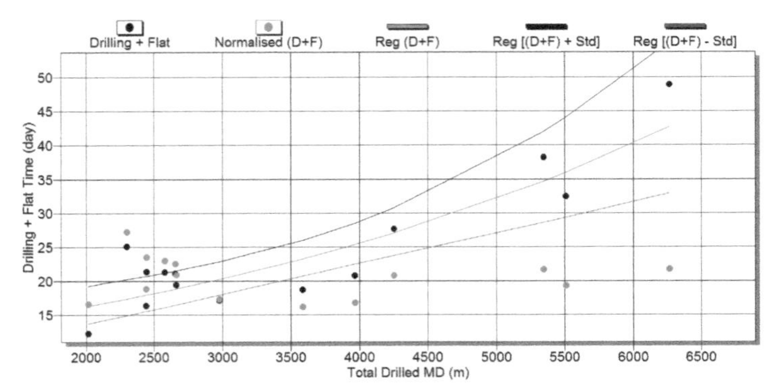

Figure 43: Wells plotted by duration of "total drilling + flat time" per drilled meterage in black dots and their depth-normalized version in red dots

Wells can be broken down in either phases or types of operations: drilling, flat time, evaluation, trouble, etc. A more precise benchmark can be done at operation level. As the operations are well defined and controlled "chunks" of time, it becomes easier to define a KPI. They basically can be assessed as duration (time) or rate (unit/time). As per duration it is quite straightforward to plot the total time of each operation, such as moving rig, working with BOP, etc, that is independent of any other factor. An example of this type of benchmark across many wells is shown in Figure 44.

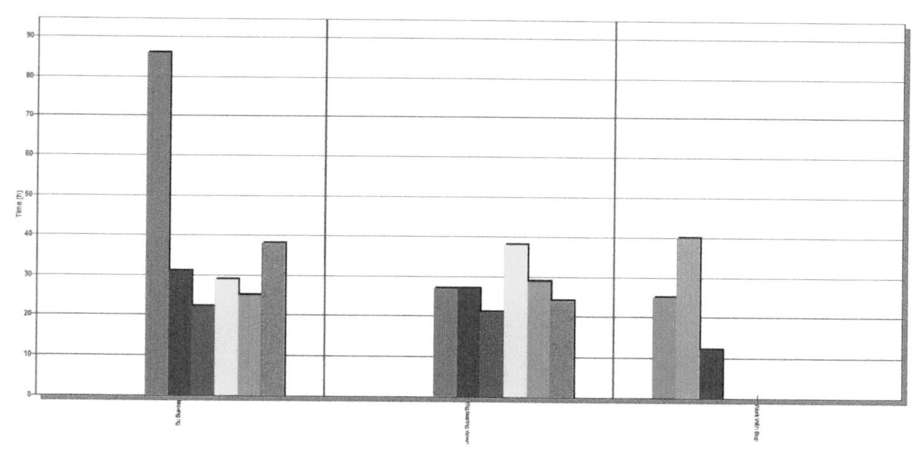

Figure 44: Single operations total time across multiple wells

A more interesting analysis is that for operations with rate or speed. Most of them are obvious, such as tripping IN and OUT operations for drill pipes or casings. Others are not so clear, such as whether "installing BOP" depends on water depth or not for offshore wells. There is a need to study some data correlation to identify if it is worth normalizing some of these operations or not. In his article Hugo Valdez (2005) shows an interesting benchmark study for operations - measured as speed - which can be the responsibility of the rig, the operator, or both. There is a need for special care when calculating speed for statistical purposes as the "time" component is the main driver for distribution, and not depth. A correct speed calculation is shown in Figure 45 where each speed operation within a well is fitted in the statistical distribution and plotted as P10, P25, P50, P75 and P90 of this distribution. Several wells can be compared side by side and with the "industry" best and worst cases to give a full picture of other performances.

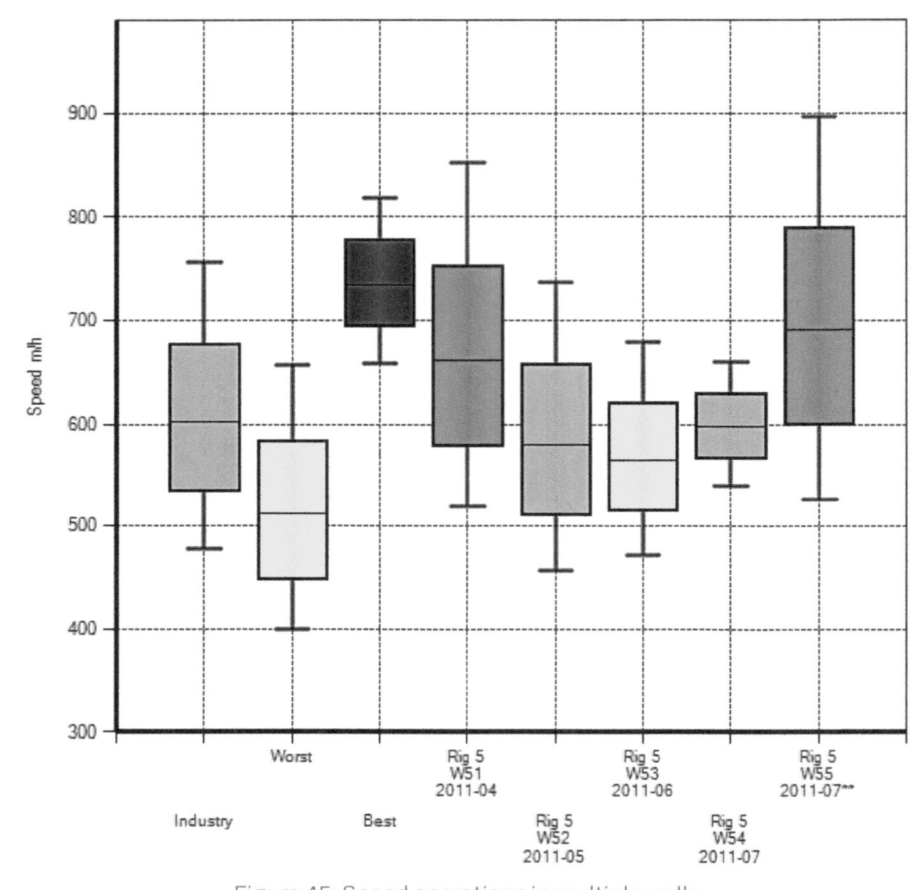

Figure 45: Speed operations in multiple wells

If the operation level is not enough, we can continue to drill down in sub-parts of each operation. With the real-time data coming from the rig it is possible to detect a "connection time" in a tripping operation. This allows benchmarking between rig crews, rig classes and other components. The article from Ketil Andersen (2009) shows how they identify and benchmark these sub-operations using rig real-time data. For example, it is possible to illustrate the spread of connection tripping speed.

As shown here, performing benchmarking can be a complex activity and it has a different meaning for different people; it can be the wrong answer for the right question. It is a minefield where we need to walk carefully. But as Peter Drucker said, "You cannot improve what you do not measure".

""What gets measured, gets managed"

Peter Drucker

6. BUSINESS CASES

The business cases presented here are based on real cases and using several techniques presented in this book. For different reasons the names and details of the companies are kept anonymous.

6.1 CASE 1 – MEASURING RESULTS

6.1.1 THE PROBLEM

A major drilling company was concerned that a comprehensive drilling database it had access to was largely unused in reducing, in any significant manner, the learning curves which would promote drilling efficiency and reduce overall drilling costs.

6.1.2 THE SOLUTION

The senior drilling management group decided to approach a data analysis company for solutions that would identify performance gaps. They wanted a solution that would also be supported by a drilling analysis tool, which could identify areas for improvement and provide the impetus to reduce overall drilling costs. They did not wish to contract consultants to do these measurements and then suggest their own improvement services. They preferred to have control over the whole process, using their own human resources, with a software tool sourced elsewhere.

6.1.3 THE IMPLEMENTATION

They purchased and implemented a drilling analysis tool, which was able to identify drilling problems by analyzing historical data.

Each phase hole was scrutinized to identify "removable time" (non-productive and invisible lost time) as the cause of the gap between the BCT and average field performance. The drilling process for each well was then broken down into phases in accordance with hole sizes. Each phase was subsequently broken into its component operations. Target times, based on the best-ever performance achieved by the company for each component operation, were established for all the operations. The pacesetter well was used to define the new planned times for each new well. Once the well was executed, post-analysis indicated the deviation from the plan and an improvements action could be suggested. With the new well, all operational times were recalculated and the new pacesetter well was continued to drive the future plans in a recursive method.

6.1.4 THE RESULT

The methodology implemented resulted in well time estimations based on optimum performance to date and the setting of benchmarks for the "perfect well" possible with current technology and operational practices. The company's calculation estimated an average well cost reduction of 15% which, when totalized, approximated CAD$7.3M for the entire drilling campaign.

6.1.5 LESSONS LEARNED

This case study shows the importance of measuring results. The success of a data management project can be calculated in terms of the return to a company's business. The hard part is to start, i.e. to know where the company is situated at a given stage in terms of performance. After that, the data analysis process is put in place and effectively implemented, so the next measurement can be done and the benefits can be measured using "before" and "after" scenarios.

6.2 CASE 2 – DRILLING MULTI-WELL CAMPAIGN

6.2.1 THE PROBLEM

A major operator had to develop a new offshore field, which presented significant challenges. As this is a completely new field it was hard to estimate the time and cost of the planned wells. They had similar historical cases from another field and the results of some exploration wells in the new field.

6.2.2 THE SOLUTION

They figured out a standard sequence to drill a new well, which required combining different sequences and types of operations from previous ones. They used the time analysis of each operation from previous wells and data from the exploration wells to estimate the approximate duration of major operations for the new wells. It was possible to extend the model and devise more complex mappings as they started drilling more wells in the new field.

6.2.3 THE IMPLEMENTATION

The implementation required a broader software application to handle multi-well campaign simulation rather than planning for only one well. Also it was necessary to adjust the scenario for each well as the historical data could not be directly used. This gave users the flexibility to design new campaigns with precise Monte Carlo simulations and be able to forecast the total cost for each campaign.

6.2.4 THE RESULT

The company incorporated the process of conceptual design of the well as the basis for the campaign in each field. The initial plan was followed up as each well was drilled, so they kept perspective of how the campaign plan was actually progressing. The planning of the entire campaign, to be developed over several years, gave the company the ability to plan resources allocation in a better way, minimizing wastage.

6.2.5 LESSONS LEARNED

Planning a detailed new well based on offset wells is a process we see often. This case study shows a much higher level of planning going into strategic development and resource allocation in several years of development, using creative data allotment.

6.3 CASE 3 – IMPLEMENT DRILLING ANALYSIS

6.3.1 THE PROBLEM

A major oil company believed their in-house "drilling analysis system" was inadequate and ineffective. They were performing drilling analysis but it had not been producing positive results. This led to frustration and a lack of confidence from the management and the data analysts. One of the major problems identified was the absence of an organized data management system as the current system was missing the acquisition of quality and formatted data that could be relied on. This resulted in drilling analysis being virtually ignored as an integral part of drilling engineering workflow. The management and the team had the will to implement, but they were just missing a good data management process and clear KPIs to improve the business.

6.3.2 THE SOLUTION

The management decided to use tools like data QC, technical limit, learning curve, and a detailed planning process. To improve their drilling performance they directed the organization into a knowledge-based structure, using already acquired drilling data, analyzed and organized using good drilling analysis tools.

6.3.3 THE IMPLEMENTATION

The company acquired a drilling analysis software package with an integrated knowledge-based methodology, centered on quality controlled historical data with a focus on a continuous learning process. Statistical analysis was used to set challenges designed to achieve technical limits based on the results of similar wells. The method provided significant indicators in the process of reducing drilling time and costs.

They also incorporated the estimations from learning curves, detailed analysis of well drilling times and problems (drilling/flat time, NPT), among other techniques, to serve as base for future wells.

A successful "drilling analysis methodology" was implemented as a systematic process to plan and follow-up wells.

6.3.4 THE RESULT

After the change to organizational processes focused on the results from the drilling analysis the company had a significant impact on overall drilling performance and costs. Similar procedures were implemented in other parts of the world where the operator drilled other wells that resulted in similar significant improvements in results.

6.3.5 LESSONS LEARNED

This case study shows that we cannot say, "drilling analysis does not work", or this is "too hard" to implement. For drilling analysis to work well, it needs to have good data and an effective process. The process also needs to start with a clear objective in mind. It is easy to get lost if you try to fix "all data" and collect everything possible. Start small with a single objective. After the success of this project, grow into others. Obviously, this depends on multiple factors in each company and the specific environment in which the company is operating.

6.4 CASE 4 – USING DATA FROM SPREADSHEETS

6.4.1 THE PROBLEM

A small oil company drilling onshore wells was using spreadsheets as the basic daily drilling reporting system. This system worked for their purposes within the company's workflow and they did not want to change it. Their problem arose when they were trying to perform some basic multi-well drilling analysis. The separate files and non-uniform data representation made it almost impossible to use the data.

6.4.2 THE SOLUTION

The company decided to get advice on data management to organize their data and implement a simple system to combine all spreadsheet files into a historical database.

6.4.3 THE IMPLEMENTATION

The spreadsheets reported by different rigs had different formats. It was decided to adopt a single format and a uniform coding system for use in future wells. The legacy data for the relevant wells was processed for incorporation into the historical database. The company contracted a specialized service that was able to automatically load every single spreadsheet into a historical database, which would enable data analytics to be performed.

6.4.4 THE RESULT

The company continues to use the spreadsheet for their drilling reporting, as it is easy to use, everyone is already familiar with it, and it has an incredibly low running cost. With the system to upload those spreadsheets into a single database they are able to follow up wells on a daily basis from the website, comparing planned vs actual, understanding operational performance, multiple wells, and other analytics that can be expected from a single database.

6.4.5 LESSONS LEARNED

This case study shows that it is not necessary to have sophisticated systems to collect data to implement comprehensive analytics. Most important is the result the company wants to achieve from the data. This is like the discussion of whether a glass is half-full or half-empty, when it is simply twice as big as it needs to be.

6.5 CASE 5 – USING COMPLETIONS DATA

6.5.1 THE PROBLEM

A major oil company sought an effective method to analyze and benchmark completions operations data to maximize data usability and consolidate diverse systems into a uniform database with common operations coding which could be re-used.

Although completions are not strictly a drilling process, this case study was incorporated here to illustrate data management in a similar process. Completions interventions are a bit harder to formalize as each intervention has a different structure and does not follow a sequence like drilling interventions do. The difficulty in this case study relates to the ability to compare different completions interventions, benchmark them against each other and plan new ones, as is done with drilling.

6.5.2 THE SOLUTION

A senior completions team utilized comprehensive analysis tools to perform time assessment from previous historical interventions. The group, aided by consultants, managed to identify and standardize common operations. As there are no obvious ways to create these standards, there were necessarily many meetings and agreements between engineers to clearly define the scope of each operation.

6.5.3 THE IMPLEMENTATION

The team first established a uniform coding system identifying all intervention types in common datasets within the same types of operations. A complete set of standard operational sequences (intervention templates) was then set to serve as benchmark reference data for future operations. The intervention template is very useful because it puts a framework in every completions sequence. The interventions are different, but the framework facilitates seeing each one under that perspective. This allows benchmarking between them and also planning for future interventions.

After the implementation was completed, each operation had specific times statistically analyzed, generating time uncertainties for each one of them. These times, with the templates created, allowed future planning and what-if scenario analysis.

6.5.4 THE RESULT

The team successfully implemented a uniform operational coding system resulting in the ability to study and compare performance between different rigs in specific fields. The study was then extended to benchmark different fields and rigs, etc, providing comprehensive analysis of performance sequences in many operational areas. The analysis also established standard benchmark formats to compare future completions operations.

The methodology implemented provided detailed (breakdown) analysis of intervention operations, identifying critical operations to be mitigated.

6.5.5 LESSONS LEARNED

This case study shows that any operational process can be broken down into identifiable operations and analyzed for benchmarking and future planning purposes.

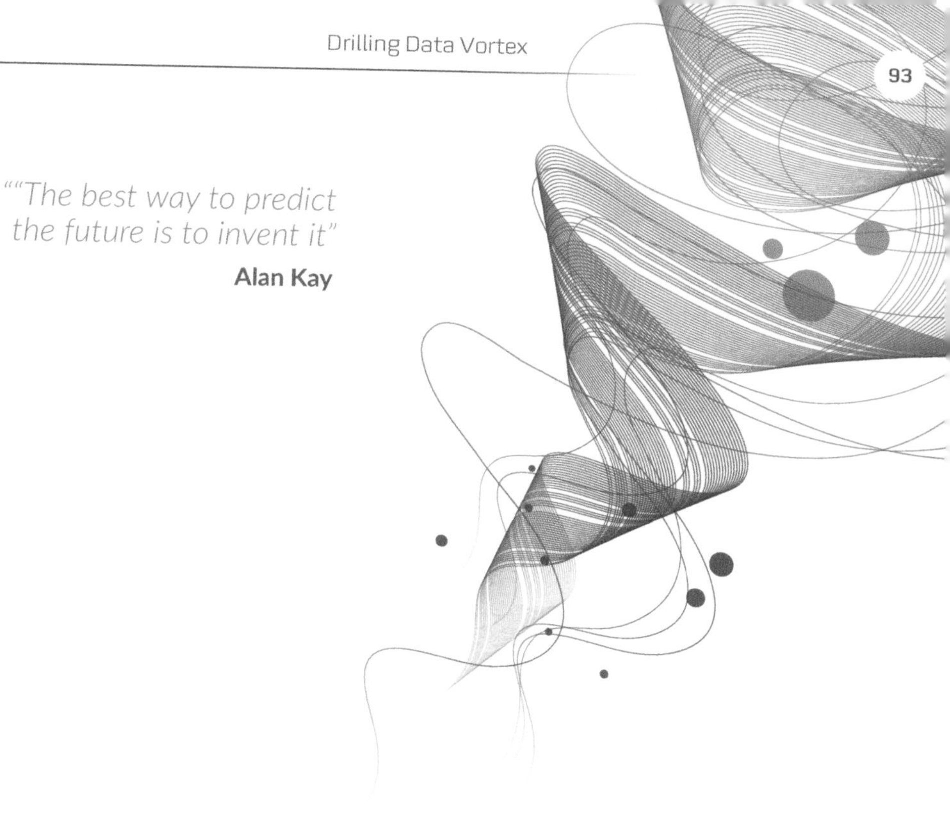

""The best way to predict the future is to invent it"

Alan Kay

7. DRILLING INTO THE FUTURE

There are many factors and issues involved in devising future directions in drilling. This chapter is confined to data management within drilling operations using the well-known model of "people, process and technology" as the cornerstone of business transformation designed to generate process improvement strategies and business process re-engineering. In general terms the industry has traditionally been heavily dependent on people: highly experienced engineers were essential for drilling the next well. It has also followed technology religiously, using the latest and greatest equipment to drill wells. The weakest leg of this tripod is process – incorporating systematic ways to collect data, analyze, plan, measure, define KPIs, etc.

In the new transformation we are witnessing now there are changes on those three fronts. "People" are changing from a strict discipline to being more generalist, as discussed in section 7.1. "Technology" is merging with other fronts, bringing "knowledge management" to a higher level, as discussed in section 7.2. Finally the "process" used is discussed in section 7.5.

7.1 THE DIGITAL DRILLING ENGINEER

Young drilling engineers are very IT savvy. They are increasingly using the types of tools described in this book, and are looking beyond drilling into the "digital oilfield". This is encapsulated in the role of digital drilling engineer (DDE), which heavily integrates engineering as a body of knowledge with business outcomes. They pursue IT-oriented tools to achieve business improvement and add value to the whole process.

Today's DDE is involved in two sets of objectives – the short and long-term. Short-term goals are achieved by creating value in running the business today. That entails improving existing process, reducing NPT, applying lessons learned, etc. The long-term target is to set the pace in employing more data and IT-oriented tools to further improve on outcomes in an ever-changing environment.

DDEs are going to be of special interest due to "big crew change" where there is a generation gap between the old experienced engineers and novice ones. Also, operations are becoming global with new development areas in difficult environments and remote locations. In these cases there is an increased need for fast communication and remote real-time data analysis, so all the expertise can be fully utilized in many sites.

The digital drilling engineer is a sub-set of the digital engineer (DE), which is well explained in the excellent book The Future Belongs to the Digital Engineer: Transforming the Industry by Dutch Holland and Jim Crompton (2013). They explain, among other issues, the importance of DE to grow "the digital oilfield" and "integrated operations" strategies. This goes beyond a specific data analysis discipline into a holistic understanding and optimization of the whole process, combining subjects such as engineering, science, IT and business processes.

As Keith Holdaway (2014) said, "The digital oilfield initiative, however, is transforming the way people work. A key ingredient of the digital oilfield is quick, easy, and timely access to quality data: Companies must shift the burden of orchestrating data from people to systems."

7.2 THE FUTURE IS NOW

Data sources need to be dynamic and recent to allow fast corrective action and decision-making. In this sense drilling data is relying more and more on real-time data. But the more data, the more data-related issues arise to acquire, QC, and analyze.

There are a number of emerging technologies that are using new data-intensive technologies as a cornerstone for better decision-making in drilling activities. They use non-standard technologies, some of them based on discoveries under the Artificial Intelligence (AI) banner. They include neural networks, fuzzy logic systems, case based reasoning (CBR), expert systems and so on. These techniques, despite their fancy names, are well-established solutions in many practical applications.

Past cases were always a source of future prediction. There is a specific technology such as CBR, which is dedicated to storing and retrieving past cases that match (usually partially) a similar context in the current situation. This mimics the human approach to solving new problems based on past experiences. Verdande Technologies (licensed to Baker Hughes) developed an innovative system called DrillEdge to predict drilling problems, such as stuck pipe, based on past "signatures", or a similar context where the problem had previously happened using real-time data. The result is shown in the "radar" plot in Figure 46.

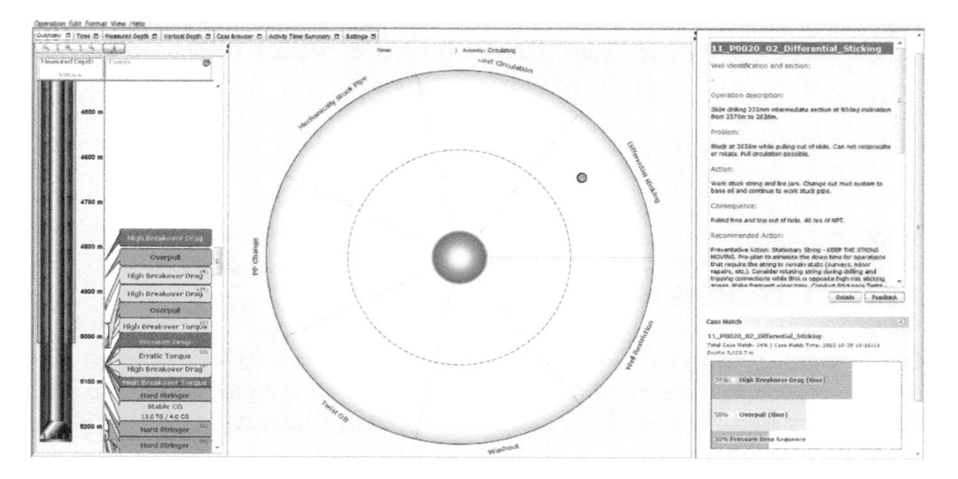

Figure 46: DrillEdge radar system. Courtesy of Verdande Technology

The reader must be familiar with dazzling Real-Time Operational Centers (RTOCs) with an endless range of features, from drilling to production data management. Constrained to drilling data, the early versions of RTOCs could only display mud-logging data and some other sensors as "raw data". It would be a job for the user to interpret the rig status from those signals. Nowadays, there are numerous solutions that can interpret these signals in real time and give more summarized and intelligent information.

They basically compress those signals in moments (Arnaout 2012) in order to combine them and recognize each operation. Although these recognitions are not perfect yet, they reach 90% accuracy and are getting better. The system ProNova can recognize those patterns and plot a full 24 hours of operations summarized in a plot, as shown in the Figure 47. Once the operations are recognized, it is possible to generate a number of plots and produce data analysis.

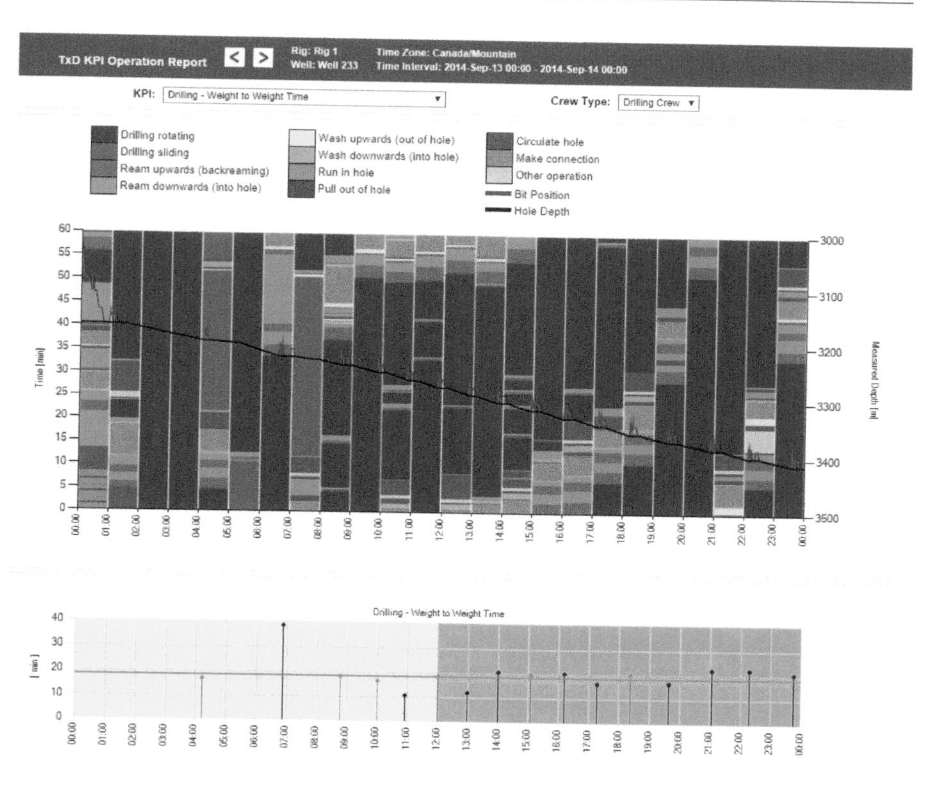

Figure 47: ProNova operation recognition system. Courtesy of TDE Group

Another interesting result is to combine these operations generated from real-time data with the report generated by people, as shown in Khudiri (2014).

This, and other solutions in the marketplace, are geared to model drilling operations as a special report. Beyond that, real-time data can be collected from geomechanical surveillance, formation evaluation, trajectory, drill bit and string behavior (vibration frequencies and energy). They can be used to run engineering modules such as wellbore stability, pore pressure prediction, torque-and-drag, drill string vibration analysis, etc. They compose a real-time drilling analyzer to allow near real-time decisions for safety and optimization. Some effort in this direction can be found in Gandelman (2010).

7.3 FORMULA RIG

There are a number of new technologies in drilling that are supported by data management and are beyond the scope of this book, which are related to engineering aspects of the drilling process. There are also many models in hydraulics, torque and drag, etc, which are not worth being detailed here. But the important aspect is that these models are now using real-time data to close the loop between what was planned and what is happening right now. This allows engineers to understand and act quickly to any variance from the expected range of operational values.

Many vendors are already providing such solutions. They make sense in RTOC where engineers can see the whole picture, in some ways better than in the rig itself. This is similar to Formula One car racing, where the team has access to all real-time data from the car and knows the car conditions better than the driver. They can send feedback with different instructions to optimize the car and help to win the race, but in the end it is the driver who is sitting in the cockpit.

This analogy leads to another concept, the automatic car, as with Google's driverless car. The developers of this car are not trying to make a Formula One car driverless, but are aiming at a new concept based on safety, stress reduction, better timing, etc. Thus, the driverless car is not the evolution of a Formula One car, but a totally new paradigm.

I am devoting time to this analogy as we have seen a lot of discussion around "drilling automation", which basically promotes less and less people on the rig until we can have a fully automatic ("driller-less") drilling rig. This is an interesting approach with consequences for safety and operations in hazardous environments. It is a difficult challenge and oil & gas companies are joining forces with other industry sectors to solve the underlying problems. If Google can provide a fully automatic car to drive on our roads, then a fully automatic rig should not be that far away. But this needs to come from another paradigm where the rigs are simpler, cheaper, and even slower than they are today. If we reduce the need to drill faster and faster, a set of current requirements may not be applicable anymore. What would be the major problem if the rig drills at half of the speed of today, but costs ¼ or ⅓ of today's costs, while totally avoiding any risks? We could have more rigs operating in parallel to compensate "time to oil" if needed. Thinking in this way turns the whole drilling business paradigm around.

A Formula One racing car serves to advance many new technologies we use in our cars nowadays. There is no question of its useful advances and innovations, but we do not drive Formula One cars everyday going from A to B. As a matter of fact, the less we need "to drive" on a daily basis, the better. In contrast, the driverless car enjoys mass transport benefits at unit level. This leads us to "factory drilling" which is today's concept for all onshore wells, especially for shale gas wells. Ideally, these rigs should just be placed in the location and then perform the rest automatically, just as a driverless car would be ready to go from "here" to "there" under no supervision. There are several initiatives in this direction, including the Drilling Systems Automation (DSA) Roadmap (http://connect.spe.org/DSARoadmap/home) with the participation of Society of Petroleum Engineers (SPE) and International Association of Drilling Contractors (IADC), among other entities.

Regardless of whether we will be "driving" a Formula One rig or a driverless rig in the near future, we will need more and more data to feed the systems – data that will mainly come from sensors in real-time. As sensors get smarter and collect more data per second, we need a bigger data depository, higher transmission rate and computing power to cope with it. In the end, it comes down to more and more "data management" and the many issues we are discussing in this book.

7.4 PROCEED WITH CAUTION

This book shows a number of data analysis techniques to improve drilling operations. Another aspect not addressed in this book is health, safety and environment (HSE). This issue is fundamental for any activities, but it involves so many aspects that it could be addressed separately in another book altogether. The increasing complexity of operations, and the number of new technologies used and "information overload", all combine to create much more stress in rig personnel who, from time to time, need to take a decision in a split second. This new (and future) scenario of more complex operations creates an unprecedented situation when it comes to the human factor. The safety culture needs to be part of the process improvement, along with all measures taken to optimize well construction. The aim should be to not only have KPIs on incidents, but also a process in place that records near-misses, incidents, accidents, etc, and a process to investigate the root causes and mitigate them in future. This is not a new process, as other industries such as aviation, the military, the nuclear industry, etc, have successfully put a safety culture in place.

Drilling companies could soon be considering drilling an offshore well in a similar fashion to flying a jumbo jet plane on a long-haul flight. To put this in perspective, the whole oil & gas industry drills approximately 80,000 wells worldwide per year, while there are about 100,000 commercial flights per day (http://www.iata.org/) transporting 3 billion passengers per year.

The aviation industry reached its great safety record by combining and sharing knowledge from airlines, pilots/crew, industry regulators, etc, with the goal of preventing accidents rather than just reacting to them. A safety culture is not only about the absence of accidents, but the entire effort to systematically avoid them on a continuous basis.

From what has already been done in terms of safety in oil & gas facilities, the near-miss analysis is of paramount importance. As William Bridges (2012) said, "Investigating near misses is critical to preventing accidents, because near-misses share the root causes of accidents; they are one or two barriers away from the actual loss/accident". It is a painstaking activity to do near-miss reports, but many industries employ it as a core activity to investigate and improve their products, services and processes. This is illustrated by the well-known Frank E. Bird triangle shown in Figure 48.

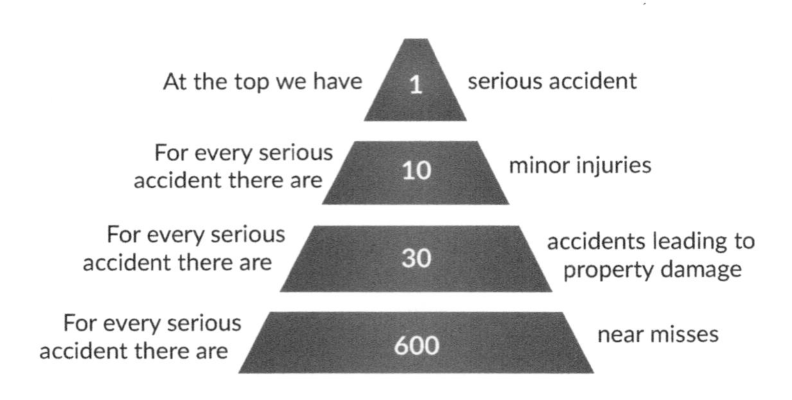

Figure 48: The accident/incident ratio by Frank E. Bird

The "cold" and rational decision made in the office by looking at the numbers and factors needs to also involve a procedure focused on how rig personnel can interact with the systems. Sometimes, there are too many alarms so most of them get ignored. IT systems and other sensors that give advice to rig personnel need to have the ability to receive feedback from users and be more adaptive to particular conditions. Systems and humans need to be better intertwined to work together effectively. For example, a given computer generated advice usually does not demonstrate the degree of uncertainty it used to come up with that decision. It should state when the underlying data used for a particular decision is knowingly unreliable (fewer data points, sensor calibration, outlier values, data variance, etc) so that, before making a decision, humans can better judge the results and the consequences. Regardless of how sophisticated the RTOCs are, with endless sensors on the rig beaming data to the office, etc, at the end of the day it is the rig personnel who are on the rig and are the first ones to suffer any HSE incident.

The comprehensive SPE report # 170575 (Committee 2014) points to some key human factors related to IT systems:

- Information overload with increasing volume of data from surface and downhole devices;
- Over-emphasis on data presentation rather than information;
- Lack of integration between data sources;
- Lack of uniform presentation of similar data or information.

Undertaking operational data analysis in the interests of process improvement is great, but it needs to be complemented by training, well-defined processes, investigation and many actions designed to reduce HSE-related issues.

7.5 THE BUSINESS OF BPM

The timeless concept of "business process management" (BPM) will be as valid tomorrow as it is today. It is soundly based on proven steps to deliver operational efficiency, along with other benefits, and to differentiate one business from its competitors. The stages of BPM may vary from author to author, but the basic concepts are shown in Figure 49.

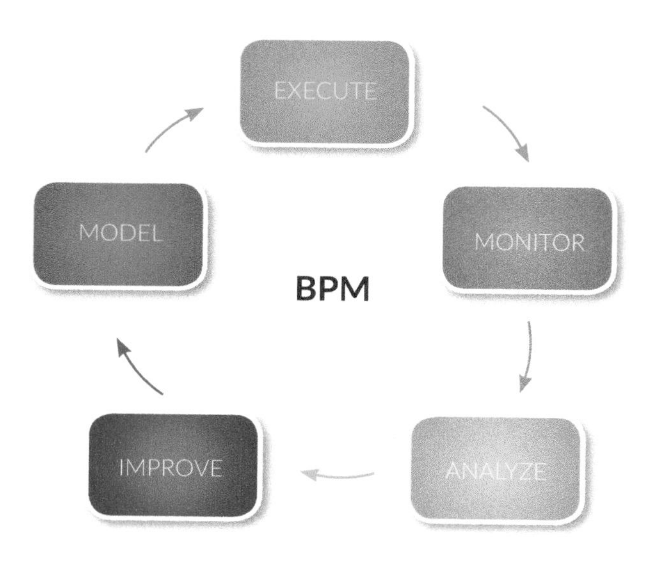

Figure 49: Business process management basic concepts

These steps are self-evident: every execution should be measured. That means in drilling, we measure via morning report or real-time data. The next stage is to analyze this data in a manner similar to that presented in Chapter 5. From these analyzes, engineers can come up with information to improve the next models (plans). This is a continuous process for improvement using business intelligence processes. The steps are well known and they just need to be fully implemented with a formal structure around them. The "process" should guarantee a better result, rather than random actions for improvement.

I would like to conclude with a quote from T.S. Eliot, along with an invitation to revisit this book, with a different understanding of drilling data management:

"We shall not cease from exploration
And the end of all our exploring
Will be to arrive where we started
And know the place for the first time."

– T.S. Eliot, Four Quartets

REFERENCES

Arnaout A., R. Fruhwirth, B. Esmael, and G. Thonhauser. 2012. „Intelligent Real-time Drilling Operations Classification Using Trend Analysis of Drilling Rig Sensors Data." Kuwait International Petroleum Conference and Exhibition. Kuwait: Society of Petroleum Engineers.

Adeleye, A.R., B.K. Virginillo, A.W. Iyoho, K. Parenteau, and H. Licis. 2004. „Improving Drilling Performance Trough Systematic Analysis of Hystorical Data: Case Study of a Canadian Field. SPE 87177." IADC/SPE Drilling Conference . Dallas, TX, USA, 2-4 March.

Andersen, Ketil , Per Arild Sjøwall, Eric Maidla, Buddy King, Gerhard Thonhauser, and Philipp Zöllner. 2009. „Case History: Automated Drilling Performance Measurement of Crews and Drilling Equipment." SPE/IADC Drilling Conference and Exhibition. Amsterdam: SPE/IADC Drilling Conference and Exhibition.

Brett, J.F., and K.. Millheim. 1986. „The Drilling Performance Curve: A Yardstick for Judging Drilling Performance. SPE 15362." 61st Annual Technical Conference and Exhibition of the Society of Pelroteum Engineers. New Orleans, LA, 5-8 October.

Bridges, William G. 2012. „Gains from Getting Near Misses Reported." 8th Global Congress on Process Safety. Houston, TX.

Committee, Technical Reports. 2014. The Human Factor: Process Safety and Culture. General, Society of Petroleum Engineers.

Few, Stephen. Now You See It. 2009. Analytics Press.

Gandelman,R.A., A.L. Martins, G.T. Teixera, A.T.A. Waldmann, A.F.L. Aragão, SPE, Petrobras, A. D. Mari, V.A. Strugata, M. S. C. Rezende, ESSS. 2010. "Real Time Drilling Data Analysis: Building Blocks for the Definition of a Problem Anticipation Methodology." SPE/IADC Drilling Conference and Exhibition. New Orleans, LA, 2-4 February.

Holland, Dutch, and Jim Crompton. 2013. The Future Belongs to the Digital Engineer: Transforming the Industry. Xlibris.

Hanschitz, Martin. 1997. AFE (Authority for Expenditure) Generation Based on Risk Analysis and Time Versus Depth Information from Similar Database Wells. Leoben: Mining University Leoben, Austria.

Hawtin, Steve. 2013. The Management of Oil Industry Exploration & Production Data. Lexington.

Holdaway, Keith. 2014. Harness Oil and Gas Big Data with Analytics: Optimize Exploration and Production with Data Driven Models. Wiley.

Hougaz, Augusto Borella, Luiz Felipe Martins, Carlos Damski, Jéssica Lima Bittencourt, and Luciano Machado Braz. 2012. „How We Improved Operations in Drilling." Rio Oil & Gas Conference. Rio de Janeiro: IBP.

Hugo Valdez, and Jurgen Sager. 2005. „Benchmarking Drilling Performance: Achieving Excellence in MODU's Operating Practices for Deepwater Drilling." SPE/IADC Drilling Conference. Amsterdam: Society of Petroleum Engineers.

Irrgang, R., C. Damski, S. Kravis, E. Maidla, and K. Millheim. 1999. „A Case-Based System to Cut Drilling Costs. SPE 56504." SPE Annual Technical Conference and Exhibition. Houston, TX, USA, 3-6 October.

Iyoho, A.W., K.K. Millheim, and M.J. Crumrine. 2005. „Lessons From Integrated Analysis of GOM Drilling Performance." SPE Drilling & Completion 6-16.

Iyoho, A.W, K.K. Millheim, B.K. Virginillo, A.R. Adeleye, and M.J. Crumrine. 2004. „Methodology and Benefits of Drilling Analysis Paradigm. SPE 87121." IADC/SPE Drilling Conference. Dallas, TX, USA, 2-4 March .

Ketil Andersen, Per Arild Sjowall, Eric Maidla, Buddy King, Gerhard Thonhauser, and Phillip Zollner. 2009. „Case History: Automated Drilling Performance Measurement of Crews and Drilling Equipment." SPE/IADC Drilling Conference and Exhibition. Amsterdam: Society of Petroleum Engineers.

Khudiri, M., J. James, M. Amer, B. Otaibi, M. Nefai, and J. Curtis. 2014. „The Integration of Drilling Sensor Real-Time Data with Drilling Reporting Data at Saudi Aramco using WITSML." Intelligent Energy Conference. Utrecht: Society of Petroleum Engineers.

Millheim, Keith, Eric Maidla, and Simon Kravis. 1998. „An Example of the Drilling Analysis Process for Extended Reach Wells. SPE 49111 ." SPE Annual Technical Conference and Exhibition. New Orleans, LA, USA, 27-30 September.

Nakagawa, Edson, Carlos Damski, and Kazuo Miura. 2005. „What is the Source of Drilling and Completions Data." Asia Pacific Oil & Gas Conference and Exhibition. Jakarta: SPE.

Negroponte, Nicholas. 1996. Being Digital. Vintage.

Rud, Olivia. 2009. Business Intelligence Success Factors: Tools for Aligning Your Business in the Global Economy. Hoboken, N.J: Wiley & Sons.

Rushmore, Peter. 2011. „Anatomy of the „Best in Class Well": How Operators Have Organised the Benchmarking of their Well Construction and Abandonment Performance." SPE/IADC Drilling Conference and Exhibition. Amsterdam: Society of Petroleum Engineers.

S. K. Peterson, J. A. Murtha, and F. F. Schneider. 1993. „Risk Analysis and Monte Carlo Simulation Applied to the Generation of Drilling AFE Estimates." SPE Annual Technical Conference and Exhibition. Houston: Society of Petroleum Engineers.

Wardt, John P. de. 2010. „Well Delivery Process: A Proven Method to Improve Value and Performance While Reducing Costs." IADC/SPE Drilling Conference and Exhibition. New Orleans: IADC/SPE

Ritu Gupta, MA (Stat), PhD, AStat,
Department of Mathematics and Statistics,
Curtin University, Australia

APPENDIX A – STATISTICS 101

A1 - PROBABILITY DISTRIBUTION

Uncertainty is an inherent characteristic of any process, natural or man-made. For instance, the amount of rainfall tomorrow, a drill bit's lifetime, the net profit of an E&P company, height of oil-water contact, total duration of a drilling intervention, types of water flooding that can be encountered during drilling, etc, are all examples of processes with uncertain outcomes. To manage process and make decisions, one needs to understand the structure of the uncertainty in the process. Probability distribution presents a systematic framework for describing and understanding uncertainty in a process.

A1.1 Random variable and probability distribution function

To describe a probability distribution, we first define the random variable as a real valued function on the sample space, taking the values on the real line R (- ,). For instance, a random variable could be different types of trouble that can be encountered in drilling. Each type of trouble (for example, stuck pipe, waiting on resources, loss of circulation) will be assigned a real number.

A random variable can be discrete or continuous. A discrete random variable can take only finite or a countable infinite set of values, for example, different types of troubles encountered in drilling, the number of tosses of a coin till you get first success. A continuous random variable can take an infinite or uncountable set of values, for example, the height of the reservoir zone in a well.

The probability distribution of a discrete random variable (X) describes the probability of the occurrence of each of its settings (p(x) = Prob(X=x)). The function p(x) is such that p(x) ≥ 0 and =1. The ordered pair (x, p(x)) is called the probability distribution of the random variable X and p(x) is referred to as a probability mass function.

Example 1: Probability distribution for types of trouble during drilling

Type of trouble during drilling (X)	Stuck pipe	Loss of circulation	Waiting on resources
Probability p(x)	0.35	0.6	0.05

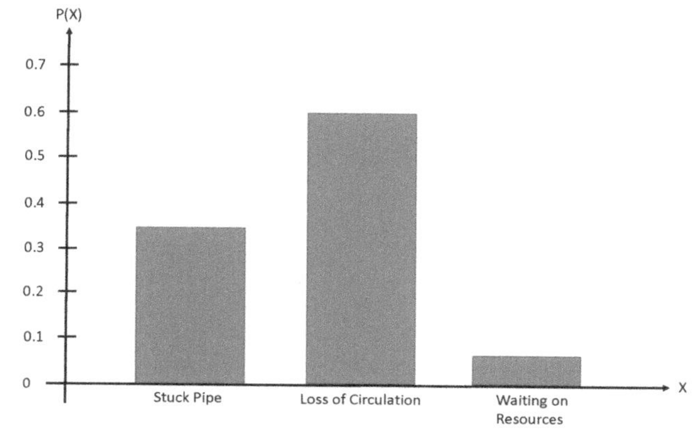

Figure 50: Probability distribution for trouble types during drilling

Similar to the discrete case, for continuous random variable (X) the probability distribution describes the probability of the occurrence of the range of values of X. The probability distribution of X is presented through a function p(x), satisfying the following conditions:

- $p(x) \geq 0$ for all values of X. It only takes positive values;
- $Prob(a \leq X \leq b) = \int_a^b p(x)dx$ (shaded area in Figure 51) is the probability that random variables take values between a and b;
- $\int_{-\infty}^{\infty} p(x)dx = 1$ the total area under the curve is 1, which means the total probability equates to 1 or 100%.

The set of all values of X for which p(x) >0 is called the support of X. The p(x) is referred to as PDF - probability density function.

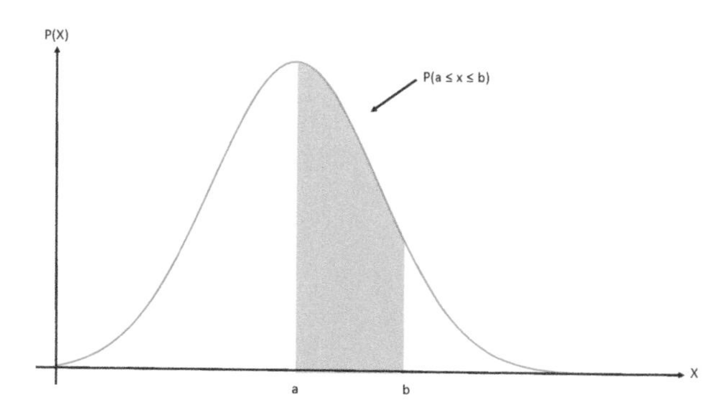

Figure 51: Probability distribution for a continuous random variable

Example 2: Probability distribution for height of oil-water contact

For example, if the height of oil-water contact (X) is equally likely to be between 10 and 30, then $p(x) = \frac{1}{20}, \; 10 < X < 30$ See Figure 52

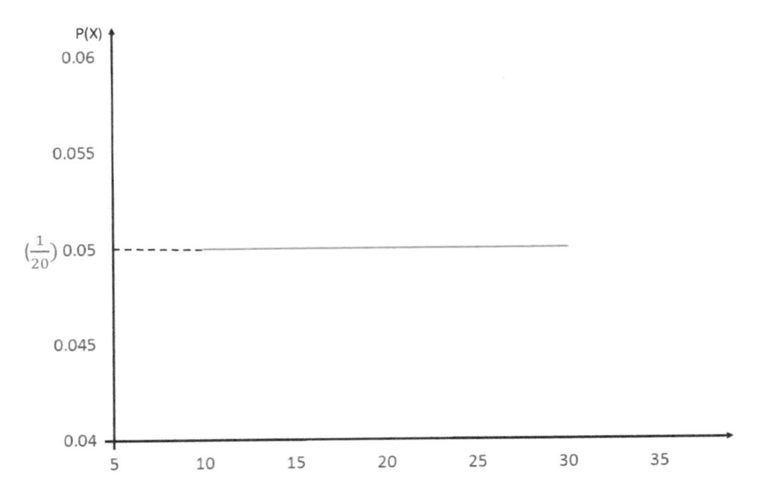

Figure 52: Distribution of height of oil-water contact

A1.2 PERCENTILES OF THE PROBABILITY DISTRIBUTIONS

The function p(x) can be used to compute probabilities for the events of interest. A quantity of usual interest in the E&P industry is P90. The P90 is the value of the random variable X, such that there is a 90% chance of realisation of the P90 value or above. Formally P90 is defined as Prob(X >P90) = 0.90 as shown in Figure 52.

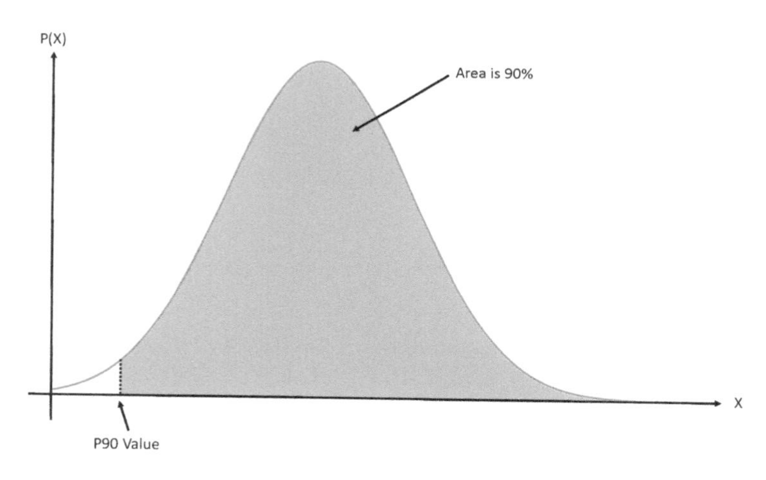

Figure 53: P90 for a probability distribution

Notice that in some documents and applications, P90 is presented as the 90% chance of realization of its value or less. Always refer to the definitions used in the context.

In the Example 2s: for random variable X, the height of oil-water contact, P90 can be calculated as:

$Prob(X > P90) = \int_{P90}^{30} p(x)dx = \frac{30-P90}{20} = 0.90$, this yields P90 =12. The P90 value is marked in Figure 54.

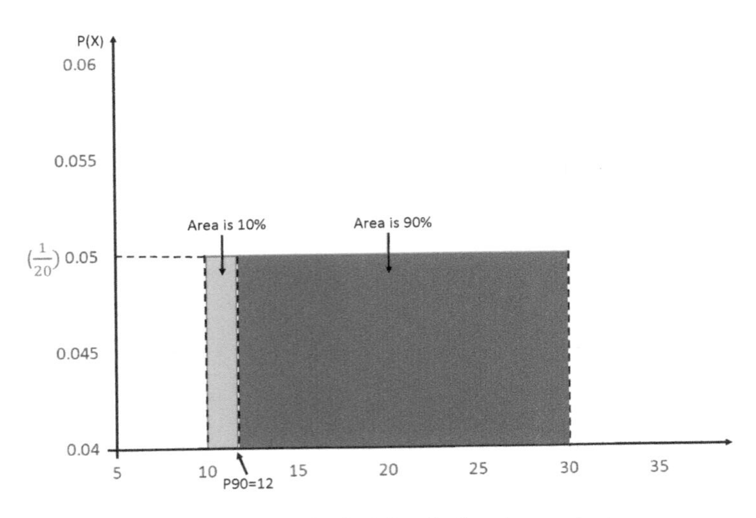

Figure 54: P90 for height of oil water contact

In general terms for any level of confidence α, Prob(X >Pα)=α as shown in Figure 55.

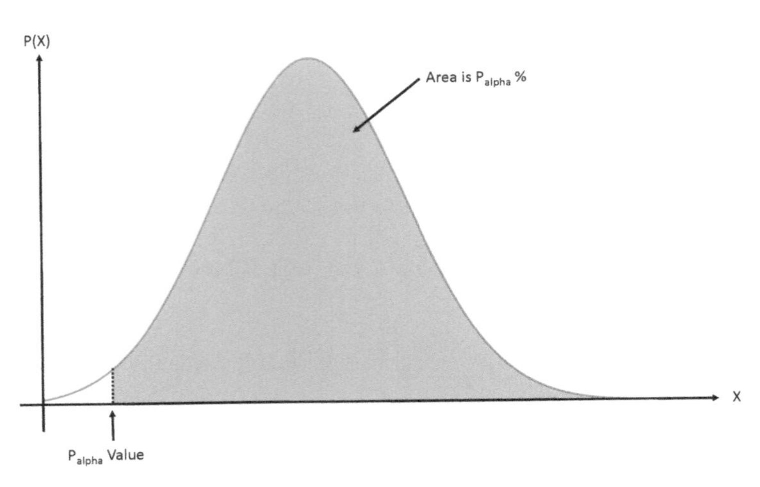

Figure 55: Percentiles at any general level

The SPE Reserves Guidelines recommend using at least three probability levels, namely P90, P50 and P10 to describe a probability function.

A1.3 SUMMARY STATISTICS FOR PROBABILITY DISTRIBUTIONS

Various summary statistics for the random variable X, like the mean and standard deviations, can be computed from the known functional form of p(x).

The mean of the distribution, usually denoted by Greek letter (μ), is the centre of gravity point of the function p(x). In this sense, mean describes the central value of the random variables X. Note that this is not always the most likely value of X.

The extent of the spread of values of X, around mean (μ), is measured by the standard deviation. Standard deviation, denoted by Greek letter Sigma (σ), is the square-root of the average of the squared distance of the values of X from μ. The square of the standard deviation is called variance.

Both the mean and standard deviation are illustrated in Figure 56.

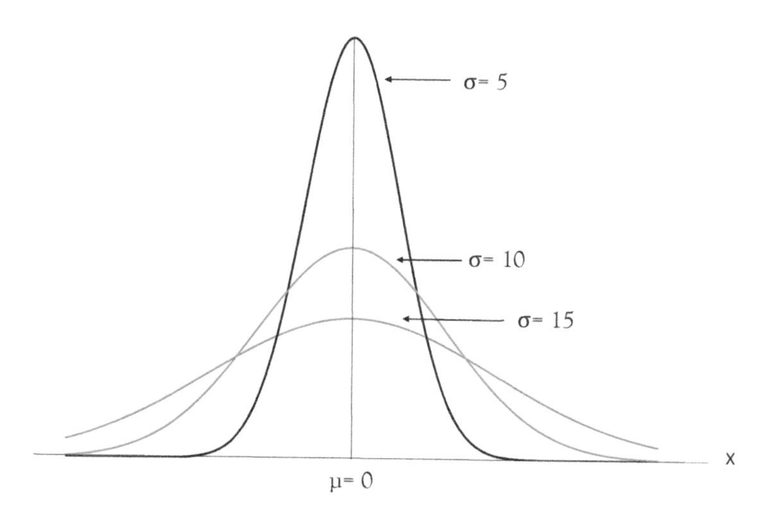

Figure 56: Three normal distribution curves with same mean and different standard deviation

Both the mean and the standard deviation are sensitive to extreme observations; hence when the probability distribution is skewed, it is better to compute the quartiles (a set of three points (Q1, Q2, Q3) that divide the total probability of X in four equal parts). The quartiles are defined as:

Prob(X <Q1)=0.25, Prob(X <Q2)=0.5 and Prob(X <Q3)=0.75. The quartiles divide the support of X into four regions referred to as quarters.

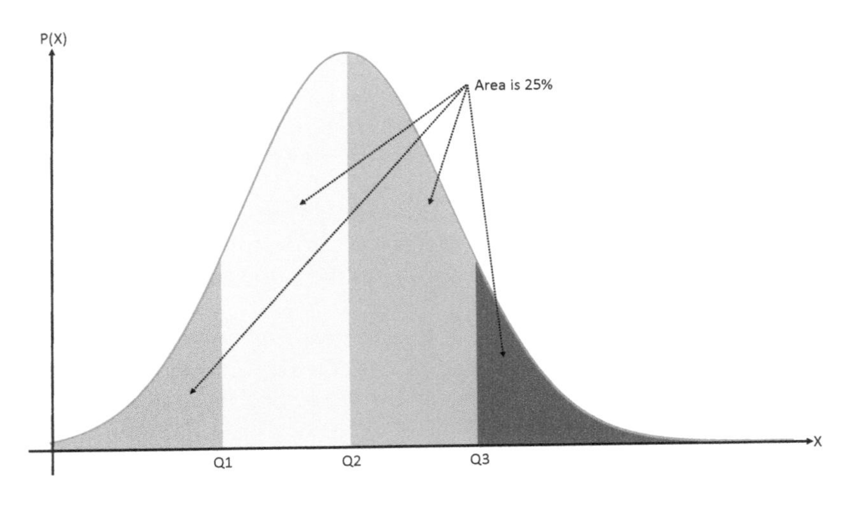

Figure 57: Quartiles for a distribution

The Q2 is also called median and is the centre point of the distribution as there is a 50% chance of getting values below and above this point. Different median values for a triangular distribution are shown in Figure 58.

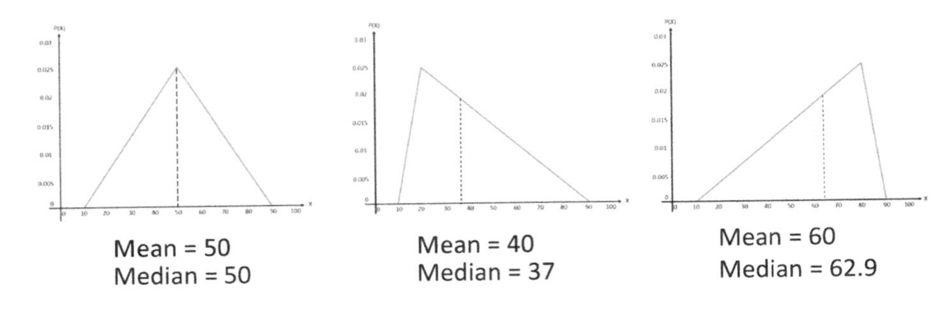

Figure 58: Comparison of mean and median for a triangular distribution

A1.4 SOME COMMON PROBABILITY DISTRIBUTIONS

The probability distribution for many processes has a common mathematical form. In this section we present three of the most commonly used probability distributions: normal, lognormal and triangular. For each probability distribution, the mathematical form and properties are presented.

A1.4.1 Normal distribution

Mathematical definition: a random variable X has normal distribution with parameter

μ and σ if $p(x) = \frac{1}{\sigma\sqrt{2\pi}} e^{-\frac{1}{2}\left(\frac{x-\mu}{\sigma}\right)^2}, -\infty < X < \infty$ as shown in Figure 59.

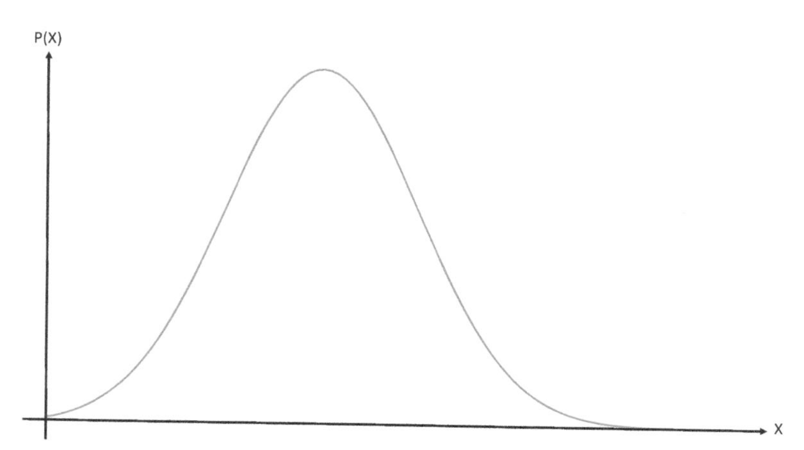

Figure 59: Probability density function for normal distribution

Settings: the distribution describes uncertainty as a symmetric bell-shaped curve, i.e. the probability is high close to the mean value and then tapers off sharply and equally as values move away from the mean.

Notation: X~N(μ, σ^2)

Properties: for this distribution, the parameters μ and σ are the mean and the standard deviation respectively. The distribution is symmetric; hence the mean and the median are equal.

Not every symmetric bell-shaped curve represents normal distribution. In a normal distribution curve (see Figure 60), there is a:

- 68.26% chance of the occurrence of values within one standard deviation of the mean. That is: $Prob(\mu - \sigma \leq X \leq \mu + \sigma) = 0.6826$
- 95.45% chance of the occurrence of values within two standard deviations of the mean. That is: $rob(\mu - 2\sigma \leq X \leq \mu + 2\sigma) = 0.9545$
- 99.74% chance of the occurrence of values within three standard deviations of the mean. That is: $Prob(\mu - 3\sigma \leq X \leq \mu + 3\sigma) = 0.9974$

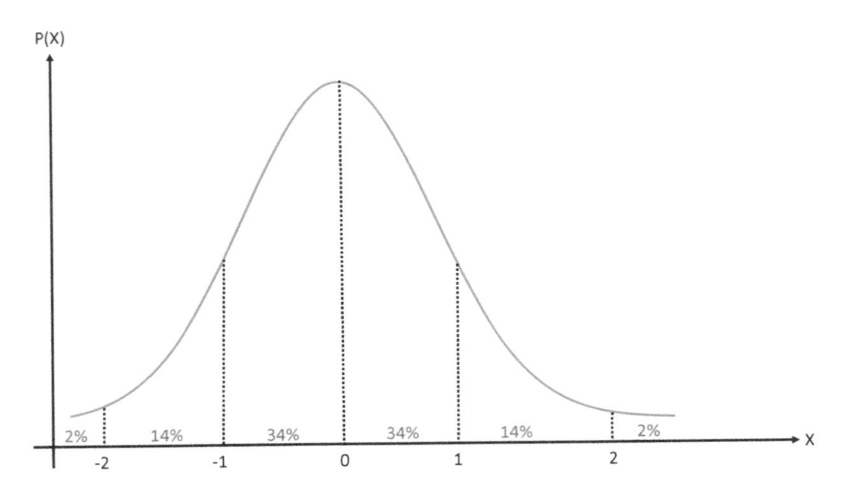

Figure 60: Area property for normal distribution (μ=0 and σ =1)

A normal distribution with values, μ=0 and σ=1 is referred to as standard normal distribution. The tables for standard normal distribution are widely available and are an integral part of even basic computing systems like Excel. These tables can be used to compute P90, P50, P10 and other quantities of interest. As normal distribution is symmetric the (P50-P90) is equal to (P50-P10), as shown in Figure 61.

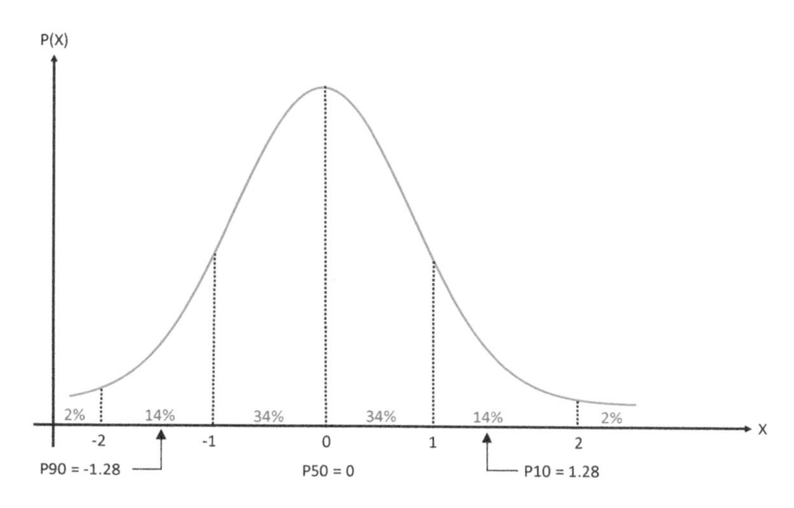

Figure 61: Percentiles for standard normal distribution

For example, let the cost of drilling a well be normally distributed with mean $320K and standard deviation of $60K. Then the parameters μ=320 and σ=60 . In this case:

- P90=243.12 [calculated in Excel using the formula "norm.inv(0.1,320,60)"], implying that there is a 90% chance that the cost of drilling will be more than $243.12K;
- P50=320, implying that there is a 50% chance that the cost of drilling will be more than $320K;
- P10=396.89 [calculated in Excel using the formula "norm.inv(0.9,320,60)"], implying that there is a 10% chance that the cost of drilling will be more than $398.89K.

Note that here P90 and P10 are equidistant from P50. For budget processing, you may be interested in calculating the probability that the cost of drilling may be more than $400K. In this case there is a 9.12% [calculated in Excel using "1 - norm.dist(400, 320, 60,True)"] chance that the cost of drilling will be more than $400K.

Application:

The normal distribution is widely used in modelling due to the central limit theorem (CLT). According to CLT, the distribution of the sample mean of a large size (sample size greater than 30) is normally distributed, whatever the shape of the underlying distribution may be (see Figure 62). Thus CLT allows you to ascertain how extreme is the sample picked for decision-making.

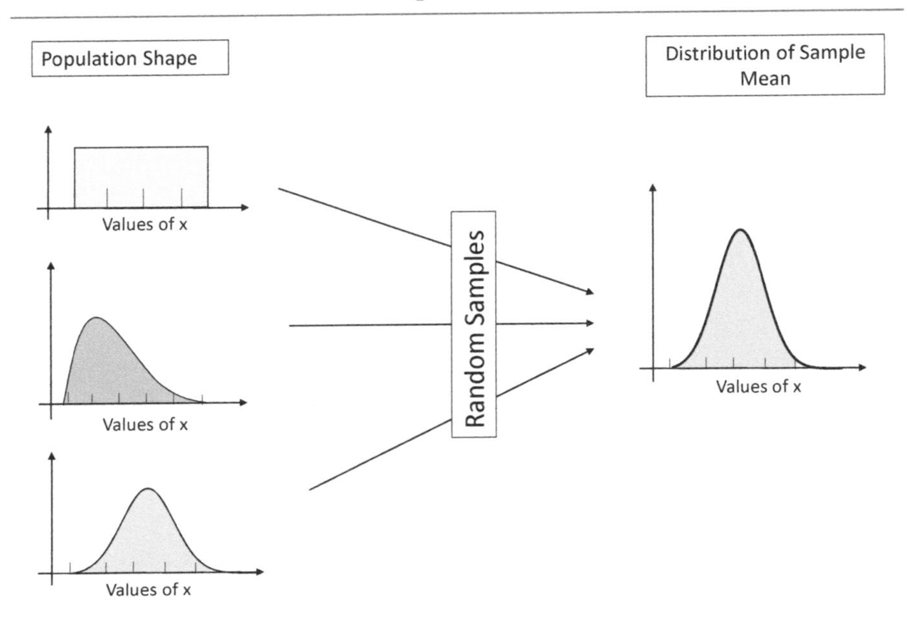

Figure 62: Central limit theorem

Many statistical tests (like the t-test) and modelling methods (linear regression) require underlying population to have normal distribution.

A1.4.2 Lognormal distribution

Definition: a random variable X has lognormal distribution with parameter μ and σ if:

$$p(x) = \frac{1}{x\sigma\sqrt{2\pi}} e^{-\frac{1}{2}\left(\frac{\log(x)-\mu}{\sigma}\right)^2}, 0 < X < \infty$$

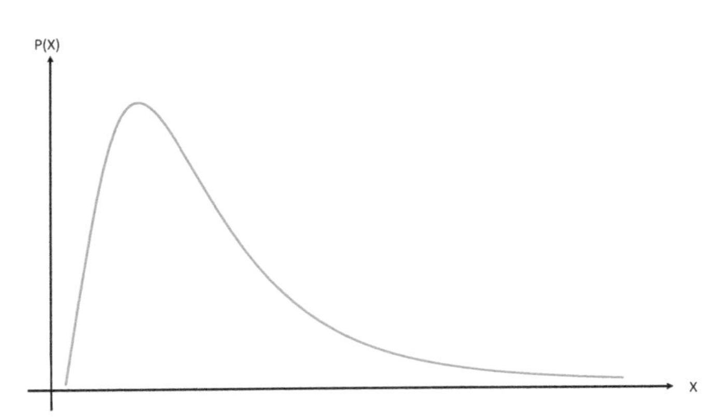

Figure 63: Probability density function of a lognormal distribution

Settings: a random variable X is said to have lognormal distribution, if Y=log(X) has a normal distribution, i.e. if we take the log() of the values of all samples in the population, those logged values have a normal distribution, as discussed earlier.

Notation: X~lognormal(μ, σ^2 (Note that μ, and σ^2 are the parameters of normal distribution for log(X)).

Properties: for this distribution the mean and standard deviations of X are:

Mean = $e^{\mu+\frac{\sigma^2}{2}}$ and

Standard deviation = $\sqrt{(e^{\sigma^2}-1)\,e^{2\mu+\sigma^2}}$.

On the log scale this distribution will possess all properties of the normal distribution. For instance, in case of lognormal distribution P90/P50=P50/P10, as log(A-B)=logA/logB. Once again, standard normal tables can be used to calculate probabilities of interest after changing the problem to the corresponding normal distribution.

Let the cost of drilling a well be lognormally distributed with mean $320K and standard deviation of $60K. Then from the formula for mean and standard deviation for lognormal distribution:

$320 = e^{\mu+\frac{\sigma^2}{2}}$ and $60 = \sqrt{(e^{\sigma^2}-1)\,e^{2\mu+\sigma^2}}$

Solving these equations lead to the parameters values as μ=5.75 and σ . In this case:

- P90=247.62 [calculated in Excel using the formula "lognorm.inv(0.1,5.75,0.1858"], implying that there is a 90% chance that the cost of drilling will be more than $247.62K;
- P50=314.19, [calculated in Excel using the formula "lognorm.inv(0.5,5.75,0.1858)"], implying that there is a 50% chance that the cost of drilling will be more than $320K;
- P10=398.66 calculated in Excel using the formula "lognorm.inv(0.9,5.75,0.1858)"], implying that there is a 10% chance that the cost of drilling will be more than $398.66K.

Note here that the ratio of P90 to P50 is the same as that of P50 to P10. Also that P50 is different from a mean of 320K.

Applications: lognormal distribution is not symmetric – it is positively skewed.

This distribution is used for modelling scenarios where you want the extreme positive observations to have a non-zero weight. For example, given a drilling operation, most of them would have none or smaller trouble time. If few of them have a larger trouble time then the distribution would be skewed; hence the lognormal distribution is more suitable to represent those operation duration samples.

A1.4.3 Triangular distribution

Definition: a random variable X has triangular distribution with parameters (a, c, b) if:

$$p(x) = \begin{cases} 2(x-a)/(b-a)(c-a), a \le x \le c \\ 2(b-x)/(b-a)(b-c), c \le x \le b. \end{cases}$$

as shown in Figure 64.

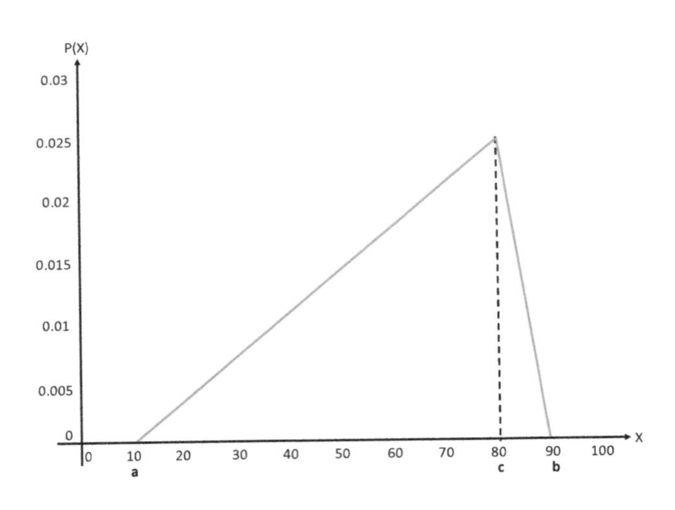

Figure 64: Probability density function for triangular distribution

Settings: the parameters a and b are the minimum and maximum values for a random variable and c is the most likely value. As we move from values a to c, the values become linearly more likely. On the same note, as we move from c to b the values on the linear scale become less likely.

Notation: X~ triangular (a, c, b).

Properties: for this distribution the mean and standard deviations of X are:

Mean = $\frac{1}{3}(a + b + c)$ and

Standard Deviation = $\sqrt{\frac{1}{18}(a^2 + b^2 + c^2 - ab - ac - bc)}$.

Applications: this distribution can take a wide variety of shapes by virtue of the choice of the most likely point c and specification of the distribution with limited knowledge.

A2 - EXPLORATORY STATISTICS

The structure of the uncertainty in any process is presented by probability distribution. In many applications the exact mathematical form of the probability distribution is not known. In such cases, to ascertain uncertainty structure a random sample is collected and exploratory statistics are computed.

The first set of exploratory statistics is about values that describe the centre tendency of the data through a single number describing a central location of the data. The sample mean X presents the average value of the sample. The centre point of the sample, called median, is such that 50% of the observations fall below and 50% above this value. The mode is the most likely value of the sample. For a given sample, the sample mean and median may not always be equal.

Once a measure of central tendency is known, it is of interest to know the spread of the sample values around the centre. Measures of dispersion like standard deviation and inter-quartile range can be computed to indicate the extent of spread of the values about centre. All of these measures can be easily computed using calculators, Excel function (average, stdev.s) or functionality available in software.

The full extent of the sample values can be described by calculating the five numbers summary: minimum value, first quartile, median, third quartile, maximum value. These numbers are graphically presented as a box plot, as shown in Figure 65.

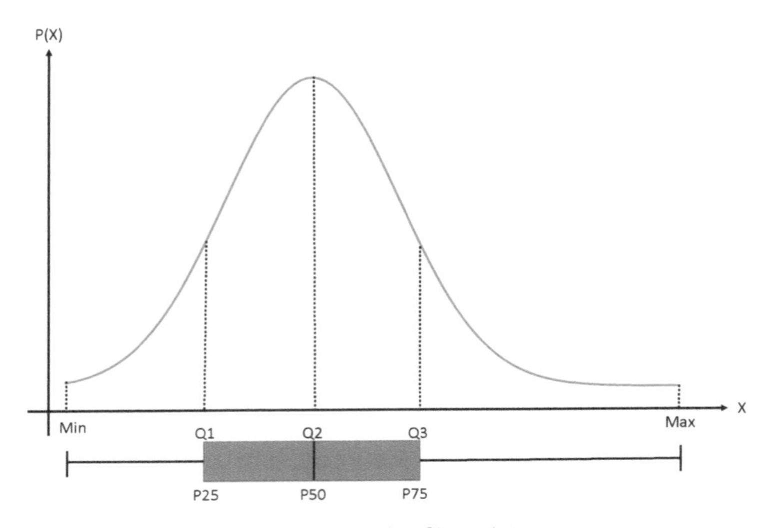

Figure 65: Example of box plot

Sometimes a sample may contain an extreme observation, very large (or small) compared to the majority of data. Such an observation is called an outlier. The easiest way to detect an outlier is to generate a box-plot using any statistical software. The outliers are presented as a star or dot beyond either end of whiskers in a box-plot, as illustrated in Figure 66. When you identify an outlier in the data, you must validate the observation for correctness. If the outlier is a valid observation, it should not be deleted. An outlier may reflect on an important possibility leading to a new discovery.

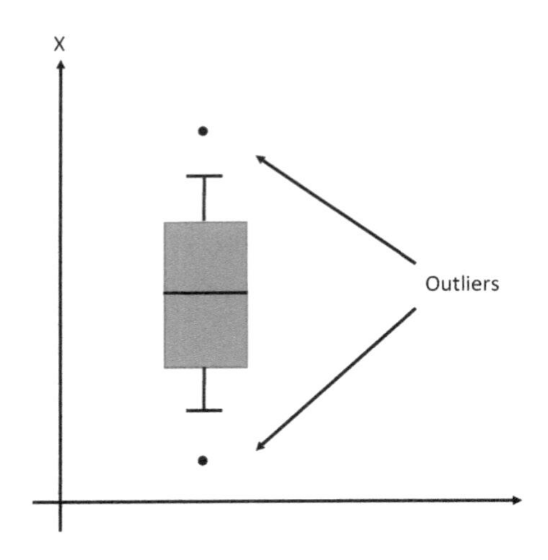

Figure 66: Outlier value in a box plot

In scenarios where the mathematical form of probability distribution is known the sample results can be used to estimate the corresponding parameters. For instance, if you pick up a random sample from normal distribution, then sample mean and standard deviation provide estimates for the parameters μ and σ. The accuracy of the estimation would depend on the sample size.

A3 - SIMULATIONS

Simulation is a powerful computing methodology, particularly useful when the exact analytical results cannot be computed or are difficult to calculate. For instance, let us consider a scenario where you plan to drill two wells. It is known that total duration (time in days) for each well is as below:

Well 1: Normally distributed with mean 90 and standard deviation of 21 days.

Well 2: Normally distributed with mean 125 and Standard deviation of 28 days.

For simplicity, let us assume that the duration for two wells is independent. From the fundamental results of normal distribution we know that the total duration of drilling (well1+well2) is also normally distributed with mean 215 and standard deviation of

$\sqrt{21^2 + 28^2} = 35.$

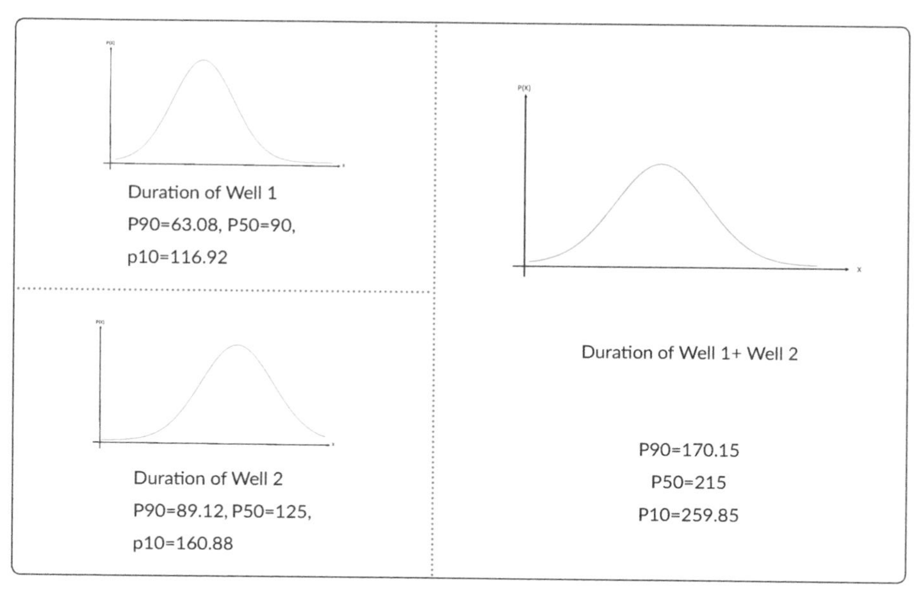

Figure 67: Analytical approach

In the above example exact analytical results are known and the uncertainty in the total duration of the well can be easily quantified. When using a simulation-based approach to solve the above problem we would be required to implement the following steps:

Step 1:

1. Randomly pick up a duration sample from well 1.
2. Randomly pick up a duration sample from well 2.
3. Add the time obtained in 1 and 2.

Step 2: Repeat step 1 a number of times (for example, 10,000).

Step 3: Collate all possible results of step 2 to create a probability distribution of total duration of drilling.

For the above example the simulation approach (see Figure 68) should give exactly the same results as in Figure 67 . There is no benefit in running simulations in this case. However, if the duration of well 1 was approximated by triangular distribution, and that of well 2 by lognormal distribution, then the total duration of the well cannot be computed analytically (this result is not known). The only way to calculate it is by using the simulation approach. This is the case when simulating a sequence of operations in a drilling campaign. Most of the operations in the sequence would have a lognormal distribution.

Run 1	Sample 1.1	RNORM(1,90,21) = 73.61
	Sample 1.2	RNORM(1,125,28) = 146.66
	Total Sample 1	73.61 + 146.66 = 220.27
Run 2	Sample 2.1	RNORM(1,90,21) = 83.21
	Sample 2.2	RNORM(1,125,28) = 139.82
	Total Sample 2	83.21 + 139.82 = 223.03
Run 3	Sample 3.1	RNORM(1,90,21) = 75.41
	Sample 3.2	RNORM(1,125,28) = 138.23
	Total Sample 3	75.41 + 138.23 = 213.64
Run N	Sample N.1	RNORM(1,90,21) = N1
	Sample N.2	RNORM(1,125,28) = N2
	Total Sample N	N1 + N2 = N

Monte Carlo Simulation Samples	
	220.27
	223.03
	213.64
	...
	N

Figure 68: Monte Carlo simulation for computing total time to drill for Figure 66 – the function RNORM is a function in R software for statistical computing for generating a random number from normal distribution with specified parameters

When performing the simulation the following principles must be followed:

1. Always allow for a large number of simulation runs (step 2 above), and observe the convergence of the results in step 3, by looking at the mean and the standard deviation of the process under study. You may use other measures like P90, P50 and P10 to monitor convergence.

2. Run multiple simulations to avoid being trapped in a single path initiated by a random seed.

3. Save the simulation seeds so that you can repeat the simulation sequence in future.

4. When simulating dependent variables (say cost of well 1 and well 2 were related), most of the software would require you to specify rank correlation for the simulation exercise. Note that rank correlation is different from Pearson correlation, and there is no one-to-one correspondence between correlation types. Check carefully whether the input required is rank or Pearson correlation.

5. When specifying correlation matrix for simulating dependent variables, ensure that it is consistent (positive definite), otherwise the system's software will auto modify the matrix.

6. To ensure uniform coverage of the uncertainty of the simulating variable, select a Latin hypercube sampling option under simulation settings.

7. Always use analytical results if available.

8. Simulation results are sensitive to the choice of your uncertainty distributions. If the exact form of the uncertainty distribution is not known, then run various simulation scenarios with different distributions.

By Jay Hollingsworth
Chief Technology Officer

APPENDIX B WITSML – AN ENERGISTICS STANDARD

B1 - BENEFITS OF STANDARDS

The need for data standards in the upstream oil & gas industry is strong. With thousands of small and large operators and national oil companies, hundreds of service and software companies and over 200 regulatory authorities in the world, custom-agreed pair-wise data transfer formats would be impossible. Larger companies and regulators may define standards they require their correspondents to follow, but that would reduce the list of potential transfer standards only slightly.

The industry has long recognized the need for some kind of widespread standards for use in exchanging information among partners, between suppliers and operators and for reporting to regulators in different countries. Organizations and technologies have sprung up in response to this need and today there are numerous organizations which provide useful standards specifically tailored to the needs of the oil & gas industry.

When data standards exist, the quality of the data as well as data volume management are greatly improved. Interoperability of multiple best of breed software applications are realized. Data integration functionality becomes seamless. Data collection for regulatory reporting is simplified. Overall, it improves efficiency, allowing personnel to focus on their jobs and spend less time dealing with data issues.

B2 - WHAT IS WITSML?

Well Information Transfer Standard Markup Language (WITSML™) is an XML data object definition and a web services specification. It is a data exchange standard that was developed to promote the seamless flow of well data between operators and service companies in addition to regulatory agencies, to speed and enhance decision-making and reporting.

The data that is transferred by WITSML consists of real-time and historical wellsite data, wellbore construction operations, operator and vendor drilling reports and well fracturing service reports.

Use of the WITSML standard data objects and data access API enables developers to create applications or it can be used to combine multiple datasets from different services companies and vendors into new applications for analysis, visualization or collaboration. Energistics has a DevKit for .NET that assists developers in implementing WITSML.

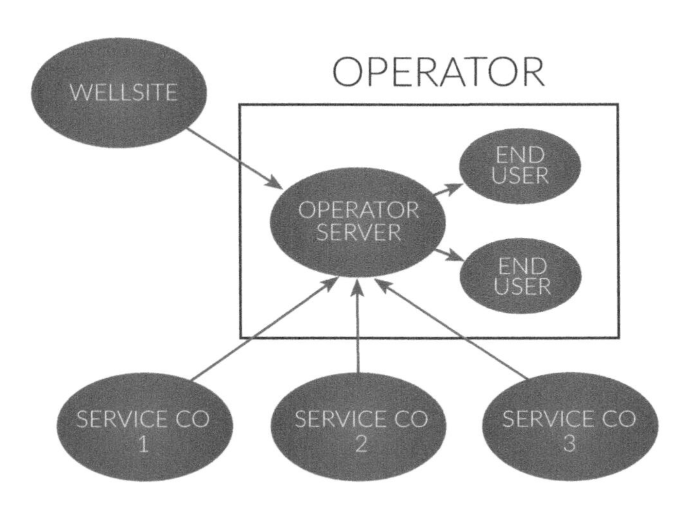

B3 - BENEFITS OF WITSML

WITSML is an important contributor to quality well construction data management and cross-discipline collaboration. It increases your capability for sharing data in real-time from platform to base to headquarters and populating geological and geophysical applications automatically. It enhances operator support and decision-making. This data is more available to future well construction teams, resulting in enhanced knowledge sharing of lessons learned.

WITSML also provides a greater ease of integrating data from different sources. The range of data sources is increased and available to be used when undertaking interpretation, optimization and delivery services on behalf of the operators. The flow of data is streamlined through internal application workflows and enables processes and applications to be used again more effectively to improve data processing efficiency.

Use of a global standard for drilling reporting simplifies data collection from operators, making it easier to add new entrants and new provinces while keeping costs lower.

Broader adoption of WITSML provides greater ease of integrating data from different sources. This increases the range of data sources that can be used when undertaking interpretation, optimization and delivery services on behalf of operators. Clear standards-based delivery from the wellsite simplifies the data gathering components when developing new solutions. This can significantly reduce the cycle time from conception to delivery.

B4 -WITSML CERTIFIED APPLICATIONS

The following industry applications have been WITSMLS V1.3.1 self-certified as of the date of this publication.

Geologix	GEO Software Suite V5.70.xx	Petris Technology	PetrisWINDS Recall Version
Geologix	Wellstore V1.00.00	PetroDAQ	Acquisition System V5
Haliburton	OpenWire V2003.0.8.1	PetroDAQ	Data Studio V1.1
HRH Gravitas	Gravitas V2.0	PetroDAQ	Drilling Control System V1
Independent Data Services	DataNet2	PetroDAQ	Drilling Guidance V1
Interactive Network Technologies	Enterprise Wellbore Viewer V3.0	PetroDAQ	Realtime Viewer V1
Knowledge Systems (now Haliburton)	Drillworks ConnectIML V1.3.1	PetroDAQ	WITSML Console V1
Kongsberg	Discovery Portal V1.2	PetroDAQ	WITSML Store Linux V1
Kongsberg	Discovery Wells V3.4	PetroDAQ	WITSML Store Windows V1
Kongsberg	Discovery Web™ V1.16	Petrolink	PowerCollect V1.0
Kongsberg	SiteCom V2007R1	Petrolink	PowerStore V1.0
LIOS Technology	DTS WITSML Data Interface V1.0	Petrolink	PowerSwitch V1.0
OFI India	WellDataNub V1.0	Petrolink	PowerView V1.0
Pason	WITSML Service V2.1	Petrolink	Saudi Aramco DRTDH V1.0
Performix (now Baker Hughes)	PetroField Genie V1.5	Polaris Guidance	Remote Drilling System V2.5
Performix (now Baker Hughes)	PetroGistics Server V1.8	Smith International	WITSML Store/ Services V1.0
Petris Technology	Petris WINDS DrillNET V1.1	Smith International	Yield Point Realtime

B5 - CASE STUDIES

US Independent oil company

An independent oil company based in the United States had drilling and completion data coming into the office from the field. The completion data had to be manually entered into an Excel spreadsheet for reporting to the headquarters office. Once the spreadsheet reached headquarters, it then had to be re-entered into the completion database. Because of all the manual effort needed to enter the data into multiple formats numerous times, the data contained lots of inconsistencies.

To solve this problem, the operator developed a WITSML adapter for Excel. Now the completion data comes in from the field and is reviewed in Excel. It then goes to the completion data all via WITSML. During this process, standard reference values are applied. All of the manual data entry is eliminated. This has greatly improved the data quality and the chain of custody has become very clear.

Saudi Aramco

At Saudi Aramco, multiple service and software companies are engaged in the drilling process. Each of the service companies has its own software infrastructure and visualization tools. Because of this, there is a lack of coherent content and no standard formats. There is no connection between the real-time data and the static master data environments.

A WITSML-based solution was implemented. Now the data enters a WITSML store from all the vendors on the project. While the data is being stored, common reference values are applied. The static master data environment was also translated into a WITSML data store. Because of this, the validated static data improved the real-time quality of the data. The integrated analysis has reduced re-work and re-keying of the data which frees up time to work on more important aspects of the job.

Pemex

Pemex had several major service and software companies as well as many smaller companies present at their drilling sites and handling their data. Each service company had its own software infrastructure and visualization tool. As a result, users were having to copy and paste or re-type the data from one vendor's application to another in order to bring the information together to perform analysis.

To reduce the manual effort involved, Pemex implemented a WITSML-based solution. The data now enters a single WITSML data store from all their vendors. Common reference values were applied during the loading process. This eliminated vendor-specific solutions, which resolved the duplicated data and improved the data quality. The manual work done to resolve the differences was eliminated.

B6 - THE FUTURE OF WITSML

The most widely-used version of WITSML is 1.3.1, but the newer version 1.4.1.1 has been in existence for over a year and adoption is accelerating. Several vendors now have WITSML servers which have been certified on 1.4.1.1 so the rate of use of the latest version is increasing.

Commercial implementations of WITSML 1.4.1.1 and earlier move data in near-real-time, meaning that the data is stored in a WITSML server and the client has to ask the server repeatedly (poll) for the latest data points.

The next generation is called WITSML 2 and it will be capable of true real-time streaming data movement via the Energistics Transport Protocol (ETP) over Websocket. This will be delivered as a Community Technology Preview to be followed by a formal release.

The Common Technical Architecture is a new technology base for WTISML 2 and the next generation of other Energistics standards. This technology base is built on other open standards to avoid licensing or other potential IP issues and to maximize the likelihood that member companies can find experienced development resources familiar with them. A list of the technologies and their uses is:

XML - eXtensible Markup Language - is the familiar technology used in the existing Energistics standards suite. Energistics currently uses XML v1.0 - updating to v1.1 would give additional flexibility in non-ASCII character sets, but there has been no request from the membership to do this (yet).

UML - Universal Modeling Language - is used internally among the membership and the Energistics staff to collaborate around a graphical depiction of a process or exchange data model. Pictures with boxes and arrows are much easier for subject matter experts to understand and comment on during the standards development process. A benefit of UML is that the XML and other transfer technologies can be generated directly from the UML pictures, which is a boost to productivity and quality (by avoiding transcription mistakes).

HDF5 - Hierarchical Data Format v5 - is a standard widely used in the scientific community for the transfer of large volumes of floating-point data. One of the downsides of XML is that a XML file can be many times larger than a simpler format because of the extra bytes associated with the self-documenting features of XML. HDF5 is close to being as small as a dataset can be, at the expense of easy human readability. But for moving seismic volumes or grids or other large datasets this tradeoff is deemed to be worthwhile.

OPC - Open Packaging Convention - is used to organize and reduce the file size of XML files involved in a transfer. In the case of an earth model a data transfer could involve hundreds of thousands of small XML documents. It would obviously be a help if these files could be organized in a folder structure and zipped to reduce the sizes of the files. There is an existing standard which does just that: the Open Packaging Convention developed by Microsoft for the files used by its Office suite (the familiar pptx, docx, xlsx, etc. files). This file format was adopted as an Ecma and ISO standard at the request of the EU, and has support directly in Visual Studio and other development toolsets.

EPC - Energistics Packaging Convention (profile of OPC) - is an Energistics add-on to OPC which adds E&P-specific file types and relationship types to the base OPC specification.

WebSocket is a standard related to html which is used to stream data over the internet as fast as TCP can carry it. In a normal internet browser session the user types a URL into the browser which identifies a web server the user wants to connect to (say, www.amazon.com/) and the resource the user would like the web server to send it (e.g., the page called index.html). This process of call and response is repeated over and over as a user chooses further links on the web page.

With Websocket, the user connects to a web server and gives the web server the identifier for a port on the user's machine and tells the web server to begin streaming data through that port as fast as it can do it. Data is streamed until the connection is broken or at the request of either party.

The Websocket protocol is being used in the next generation of Energistics standards to support true real-time streaming of data from a server to a client, whether the server is a drilling rig, a producing well, or a database serving data to a series of clients.

Avro – When communicating in a spoken language, sentences are recited one word at a time because we are unable to say multiple words simultaneously. Serial data protocols like Websocket have the same limitation. In order for the bytes of data to be transmitted over a wire, they have to be broken up into bytes and each byte broken into bits. There has to be a mirror-image set of processes which break up the data into a known sequence to put on the wire and then reassemble the data into meaningful structures at the receiving end. This process is broadly called serialization.

Avro is a data serializer developed and maintained by the Apache open source community. Avro has the interesting feature that the structure of the data doesn't have to be decided in advance – it simply informs the listening client of the structure of the data it is about to receive. This gives upstream oil & gas the flexibility it needs to be able to make changes at a rig site or producing well dynamically.

JSON – Javascript Object notation is used by Avro to describe the structure of the incoming data stream.

ETP – Energistics Transfer Protocol is the name given to a new standard protocol which uses Websocket, Avro and JSON to transfer real-time and static data from a server to a client. The protocol is a simple API consisting of messages passed between the client and server to initiate and close sessions, identify the data available on a server, initiate transfer of some subset of that data, and some other functions. ETP will eventually replace the existing APIs in all the Energistics standards.

Standard naming and other conventions – The final component of the Common Technical Architecture is the documenting of a standard set of naming and other conventions plus units of measure used across all the Energistics standards. Much of this work is not new, but is derived from prior art in Energistics and from widely-respected development shops which are willing to publicize their internal coding standards, like Microsoft and Google.

B7 - OTHER ENERGISTICS STANDARDS

Energistics is a global, not-for-profit, membership consortium that serves as the facilitator, custodian and advocate for the development and adoption of technical, open data exchange standards for the upstream oil and gas industry. Energistics provides a neutral collaboration environment and anti-trust framework for the ongoing development of open data exchange standards. In addition to WITSML, Energistics has two other flagship standards: PRODML™ and RESQML™.

PRODML supports data exchange representing the flow of fluids from the point they enter the wellbore to the point of custody transfer, together with the results of the field services and engineering analyzes required in production operation workflows, in a vendor-neutral, open format. PRODML serves to improve reporting processes and production data exchange between 3rd parties. PRODML reduces complexity, implementation time, risk in application integration and costs. It also promotes efficiency, compliance, data quality and innovation.

RESQML is the data exchange format for transferring earth model data between applications in a vendor-neutral, open and simple format. This standard provides a structural and stratigraphic framework comprised of horizons and faults, geobodies, fractures, fluid contact, unconformity surfaces, and the chronological relationships between them, in time and depth. RESQML improves grid capabilities to support unlimited numbers of multi-million cells in a model and provisioning for unstructured grids. It also provides georeferencing and easy transfers to GIS-aware databases.

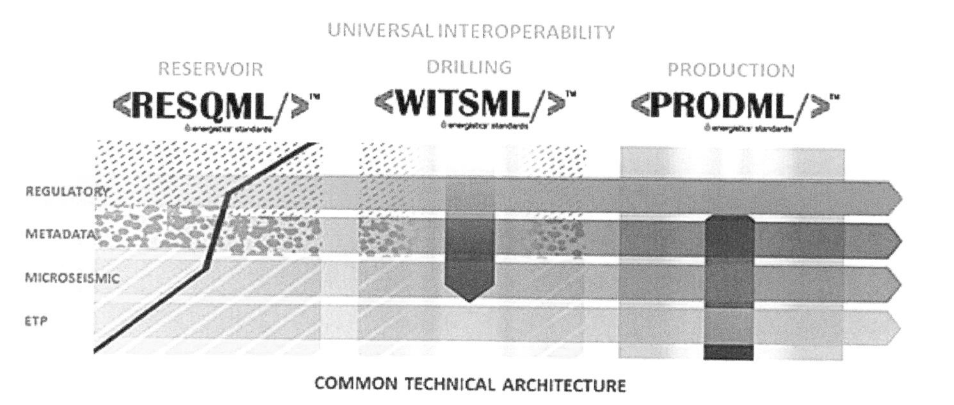

APPENDIX C – COPYRIGHT NOTICE

Throughout this book several product names were used to illustrate the subject. This book does not intend to be a complete catalogue of products. The author did not receive any request to include or exclude product names from this text. The product names are the copyright of each vendor and it is their responsibility to fulfill the features described in this book. The author's sole purpose is convey possible features resulting from data management and use examples to partially illustrate those features.

Product Name	Owner
Business Objects	SAP AG
Cognos	IBM
DataVera	Petris/Halliburton
DrillEdge	Verdande Technology
Genesis iVAC	Genesis Petroleum Technologies
Genesis Xcheck	Genesis Petroleum Technologies
Genesis Xplot	Genesis Petroleum Technologies
Hyperion	Oracle
OpenSpirit	TIBCO Software
Oracle	Oracle
ProNova	TDE
Spotfire	TIBCO
Windows, MSSQL Server, MS Access, MS Excel	Microsoft
WITSML	Energistics

CPSIA information can be obtained at www.ICGtesting.com
Printed in the USA
LVOW02s1212221214

3827LVUK00011B/51/P

9 780994 164209